As affecting the fate of my absent husband

MCGILL-QUEEN'S NATIVE AND NORTHERN SERIES
(In memory of Bruce G. Trigger)

Sarah Carter and Arthur J. Ray, Editors

1 *When the Whalers Were Up North*
Inuit Memories from the Eastern Arctic
Dorothy Harley Eber

2 *The Challenge of Arctic Shipping*
Science, Environmental Assessment, and Human Values
Edited by David L. VanderZwaag and Cynthia Lamson

3 *Lost Harvests*
Prairie Indian Reserve Farmers and Government Policy
Sarah Carter

4 *Native Liberty, Crown Sovereignty*
The Existing Aboriginal Right of Self-Government in Canada
Bruce Clark

5 *Unravelling the Franklin Mystery*
Inuit Testimony
David C. Woodman

6 *Otter Skins, Boston Ships, and China Goods*
The Maritime Fur Trade of the Northwest Coast, 1785–1841
James R. Gibson

7 *From Wooden Ploughs to Welfare*
The Story of the Western Reserves
Helen Buckley

8 *In Business for Ourselves*
Northern Entrepreneurs
Wanda A. Wuttunee

9 *For an Amerindian Autohistory*
An Essay on the Foundations of a Social Ethic
Georges E. Sioui

10 *Strangers Among Us*
David Woodman

11 *When the North Was Red*
Aboriginal Education in Soviet Siberia
Dennis A. Bartels and Alice L. Bartels

12 *From Talking Chiefs to a Native Corporate Elite*
The Birth of Class and Nationalism among Canadian Inuit
Marybelle Mitchell

13 *Cold Comfort*
My Love Affair with the Arctic
Graham W. Rowley

14 *The True Spirit and Original Intent of Treaty 7*
Treaty 7 Elders and Tribal Council with Walter Hildebrandt, Dorothy First Rider, and Sarah Carter

15 *This Distant and Unsurveyed Country*
A Woman's Winter at Baffin Island, 1857–1858
W. Gillies Ross

16 *Images of Justice*
Dorothy Harley Eber

17 *Capturing Women*
The Manipulation of Cultural Imagery in Canada's Prairie West
Sarah A. Carter

18 *Social and Environmental Impacts of the James Bay Hydroelectric Project*
Edited by James F. Hornig

19 *Saqiyuq*
Stories from the Lives of Three Inuit Women
Nancy Wachowich in collaboration with Apphia Agalakti Awa, Rhoda Kaukjak Katsak, and Sandra Pikujak Katsak

20 *Justice in Paradise*
Bruce Clark

21 *Aboriginal Rights and Self-Government*
The Canadian and Mexican Experience in North American Perspective
Edited by Curtis Cook and Juan D. Lindau

22 *Harvest of Souls*
The Jesuit Missions and Colonialism in North America, 1632–1650
Carole Blackburn

23 *Bounty and Benevolence*
A History of Saskatchewan Treaties
Arthur J. Ray, Jim Miller, and Frank Tough

24 *The People of Denendeh*
Ethnohistory of the Indians of Canada's Northwest Territories
June Helm

25 *The Marshall Decision and Native Rights*
Ken Coates

26 *The Flying Tiger*
Women Shamans and Storytellers of the Amur
Kira Van Deusen

27 *Alone in Silence*
European Women in the Canadian North before 1940
Barbara E. Kelcey

28 *The Arctic Voyages of Martin Frobisher*
An Elizabethan Adventure
Robert McGhee

29 *Northern Experience and the Myths of Canadian Culture*
Renée Hulan

30 *The White Man's Gonna Getcha*
The Colonial Challenge to the Crees in Quebec
Toby Morantz

31 *The Heavens Are Changing*
Nineteenth-Century Protestant Missions and Tsimshian Christianity
Susan Neylan

32 *Arctic Migrants/Arctic Villagers*
The Transformation of Inuit Settlement in the Central Arctic
David Damas

33 *Arctic Justice*
On Trial for Murder – Pond Inlet, 1923
Shelagh D. Grant

34 *Eighteenth-Century Naturalists of Hudson Bay*
Stuart Houston, Tim Ball, and Mary Houston

35 *The American Empire and the Fourth World*
Anthony J. Hall

36 *Uqalurait*
An Oral History of Nunavut
Compiled and edited by John Bennett and Susan Rowley

37 *Living Rhythms*
Lessons in Aboriginal Economic Resilience and Vision
Wanda Wuttunee

38 *The Making of an Explorer*
George Hubert Wilkins and the Canadian Arctic Expedition, 1913–1916
Stuart E. Jenness

39 *Chee Chee*
A Study of Aboriginal Suicide
Alvin Evans

40 *Strange Things Done*
Murder in Yukon History
Ken S. Coates and William R. Morrison

41 *Healing through Art*
Ritualized Space and Cree Identity
Nadia Ferrara

42 *Coyote and Raven Go Canoeing*
Coming Home to the Village
Peter Cole

43 *Something New in the Air*
The Story of First Peoples Television Broadcasting in Canada
Lorna Roth

44 *Listening to Old Woman Speak*
Natives and Alternatives in Canadian Literature
Laura Smyth Groening

45 *Robert and Francis Flaherty*
A Documentary Life, 1883–1922
Robert J. Christopher

46 *Talking in Context*
Language and Identity in Kwakwaka'wakw Society
Anne Marie Goodfellow

47 *Tecumseh's Bones*
Guy St-Denis

48 *Constructing Colonial Discourse*
Captain Cook at Nootka Sound
Noel Elizabeth Currie

49 *The Hollow Tree*
Fighting Addiction with Traditional Healing
Herb Nabigon

50 *The Return of Caribou to Ungava*
A.T. Bergerud, Stuart Luttich, and Lodewijk Camps

51 *Firekeepers of the Twenty-First Century*
First Nations Women Chiefs
Cora J. Voyageur

52 *Isuma*
Inuit Video Art
Michael Robert Evans

53 *Outside Looking In*
Viewing First Nations Peoples in Canadian Dramatic Television Series
Mary Jane Miller

54 *Kiviuq*
An Inuit Hero and His Siberian Cousins
Kira Van Deusen

55 *Native Peoples and Water Rights*
Irrigation, Dams, and the Law in Western Canada
Kenichi Matsui

56 *The Rediscovered Self*
Indigenous Identity and Cultural Justice
Ronald Niezen

57 *As affecting the fate of my absent husband*
Selected Letters of Lady Franklin Concerning the Search for the Lost Franklin Expedition, 1848–1860
Edited by Erika Behrisch Elce

As affecting the fate of my absent husband

Selected Letters of Lady Franklin Concerning the Search for the Lost Franklin Expedition, 1848–1860

EDITED, WITH AN INTRODUCTION
AND NOTES, BY ERIKA BEHRISCH ELCE

McGill-Queen's University Press
Montreal & Kingston · London · Ithaca

ISBN 978-0-7735-3479-7

Legal deposit first quarter 2009
Bibliothèque nationale du Québec

Printed in Canada on acid-free paper that is 100% ancient forest free (100% post-consumer recycled), processed chlorine free

McGill-Queen's University Press acknowledges the support of the Canada Council for the Arts for our publishing program. We also acknowledge the financial support of the Government of Canada through the Book Publishing Industry Development Program (BPIDP) for our publishing activities.

Frontispiece: Portrait of Lady Franklin by Thomas Bock, c. 1840. Crayon and pencil, 34.0 cm x 30.7 cm. Queen Victoria Museum and Art Gallery, Launceston, Tasmania, Australia, QVM FA 497.

LIBRARY AND ARCHIVES CANADA CATALOGUING IN PUBLICATION

Franklin, Jane, Lady, 1792–1875.
As affecting the fate of my absent husband : selected letters of Lady Franklin concerning the search for the lost Franklin expedition, 1848–1860 / edited, with an introduction and notes, by Erika Behrisch Elce.

Includes bibliographical references and index.
ISBN 978-0-7735-3479-7

1. Franklin, Jane, Lady, 1792–1875 – Correspondence. 2. Franklin, John, Sir, 1786–1847. 3. Search and rescue operations – Arctic regions. 4. Search and rescue operations – Northwest Territories. 5. Arctic regions – Discovery and exploration – British. 6. Explorers' spouses – Great Britain – Correspondence. I. Elce, Erika Behrisch II. Title.

G662.F728 2009 910.9163'27 C2008-905655-8

Set in 10.5/14 Minion and 11.5/14 Mrs Eaves with Dalliance Flourishes
Book design and typesetting by Garet Markvoort, zijn digital

To my grandmothers,
Erika Hilda Behrisch and Ruth Ellen Beauchamp

CONTENTS

ACKNOWLEDGMENTS

The road to completing this book has been long and winding, and I am happy to thank both institutions and people for their support along the way. I am indebted to Queen's University at Kingston, Ontario, for its generous financial support during the research stage of this project, and to the Royal Military College of Canada for the time it allowed me to finish the manuscript. Roger Martin of McGill-Queen's University Press provided advice and encouragement, helping me keep my spirits up. As with all peer-reviewed works, so much is owed to the anonymous reviewers who read and provided useful responses to the manuscript. I am also fortunate to be able to offer heartfelt thanks to Dr C. Stuart Houston, whose enthusiasm for the project – and whose myriad constructive suggestions – gave me the push I needed to persevere. Dr Houston's work on Arctic literature and exploration has been an inspiration to me, and I am honoured to have had him in my corner. Mel Wiebe, Mary Millar, and Ellen Hawman of the Disraeli Project (Queen's University) were my initial and constant cheerleaders and mentors; without their early and continued support this book would never have come to fruition. Michael Pacey produced beautiful maps for the book, and I am grateful to the National Maritime Museum (London), the Scott Polar Research Institute (Cambridge), and the Queen Victoria Museum and Art Gallery (Tasmania) for permission to use images from their respective collections. Finally, thank you to the members of my family, who were happy I had a project on the go and are equally happy it's done.

ABBREVIATIONS

ANB	*American National Biography*
BDL	*Benjamin Disraeli Letters*
DAB	*Dictionary of American Biography*
DCB	*Dictionary of Canadian Biography*
DNB	*Dictionary of National Biography*
DP	Derby Papers
DPBK	*Debrett's Peerage, Baronetage, and Knightage*
EA	*Encyclopædia Americana*
EB	*Encyclopædia Britannica*
G	*Gazetteer of the British Isles*
H	Hughenden Archives
HBC	Hudson's Bay Company
HBD	*Haydn's Book of Dignities*
HBL	Peter Cunningham's *Hand Book of London*
HCSP	*House of Commons Sessional Papers*
HLSP	*House of Lords Sessional Papers*
HPD	*Hansard's Parliamentary Debates*
MEB	*Modern English Biography*
NLC	National Library of Canada
NMMC	*National Maritime Museum Catalogue*
POD	*London Post Office Directory*
PRO	Public Record Office
SPRI	Scott Polar Research Institute
RGS	Royal Geographical Society

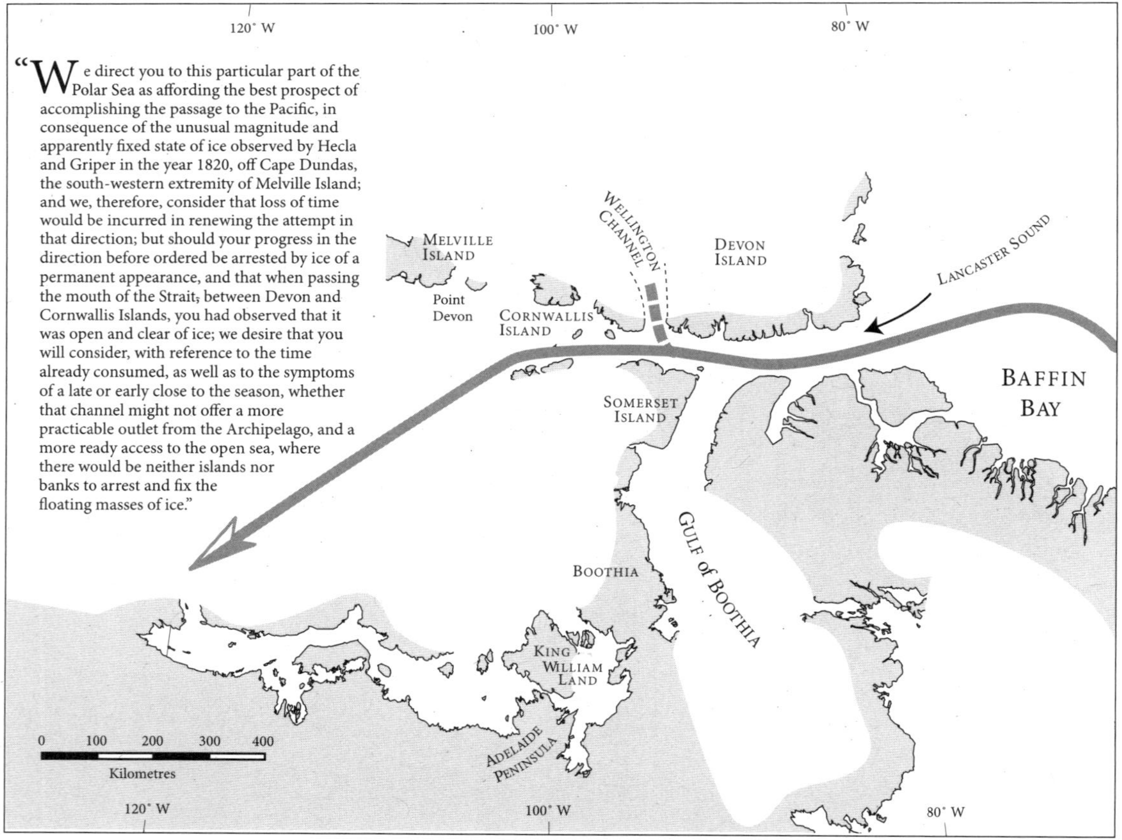

Map 1
Franklin's instructions and intended route, 1845

Map by Michael Pacey

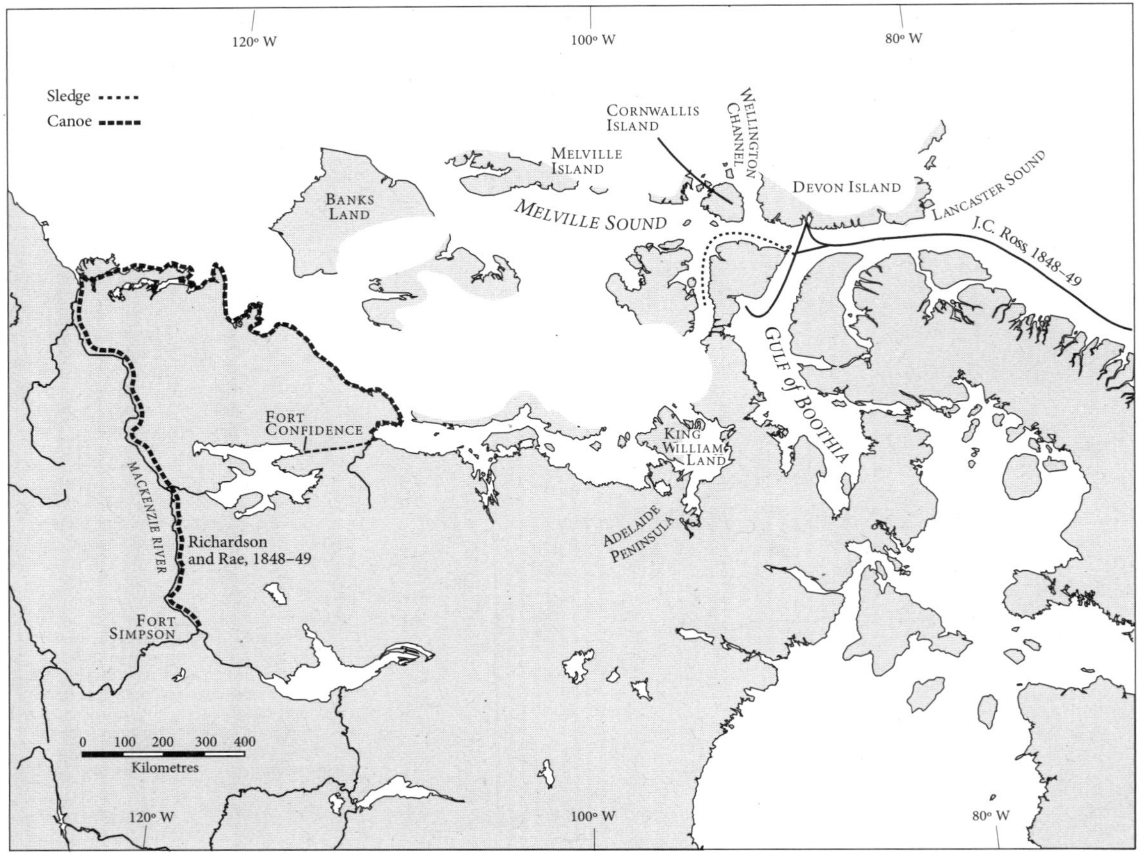

Map 2
Expeditions,
1847–1849

Map by Michael Pacey

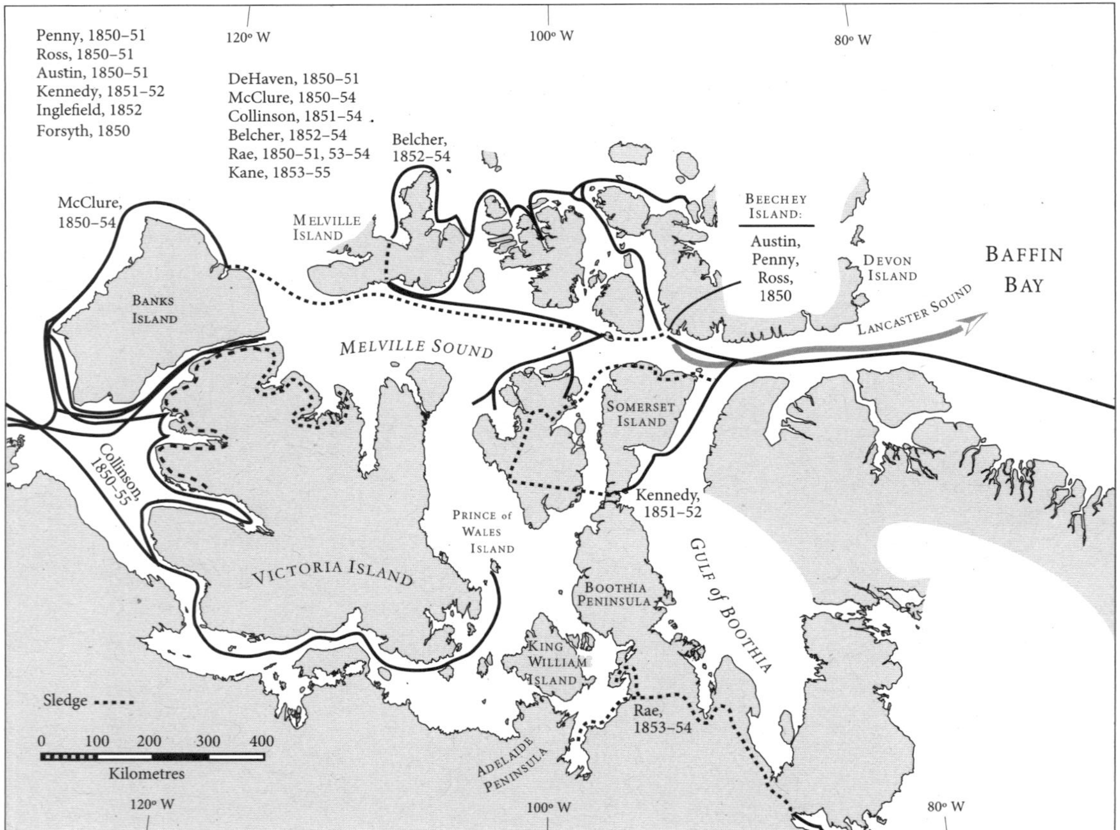

Map 3
Expeditions,
1850–1855

Map by Michael Pacey

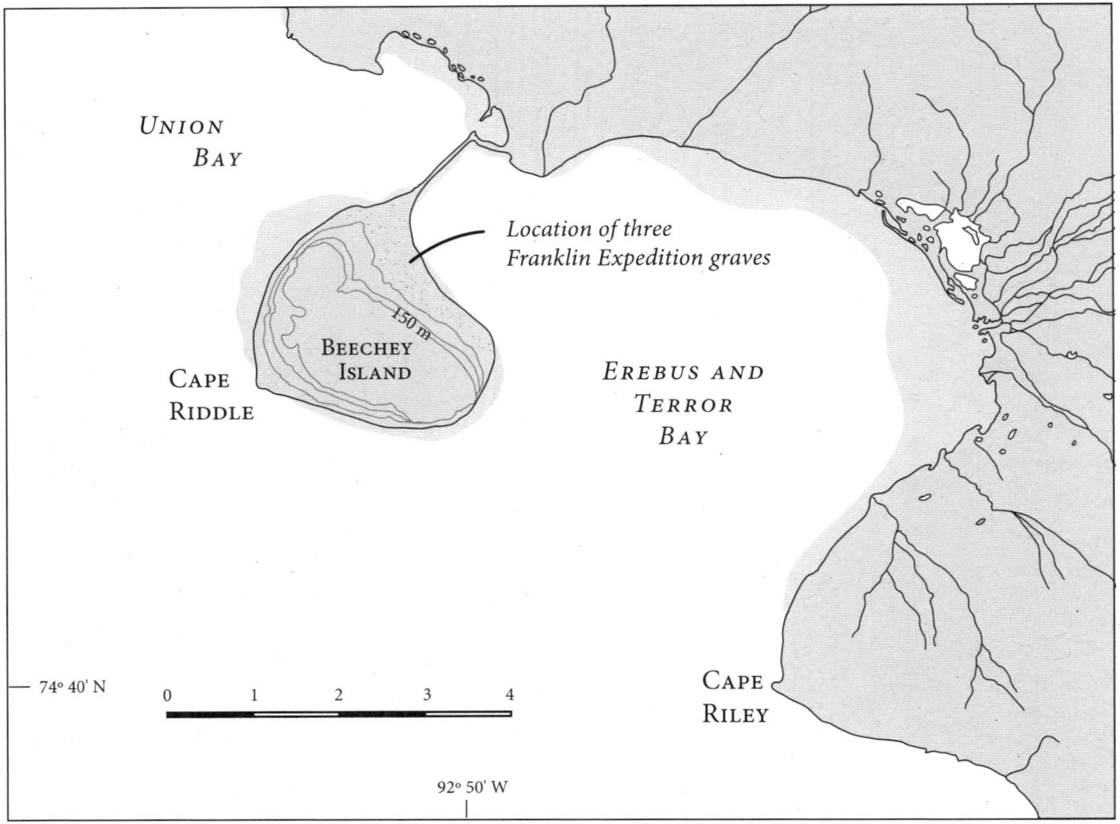

Map 4
Beechey Island
and Cape Riley

Map by Michael Pacey

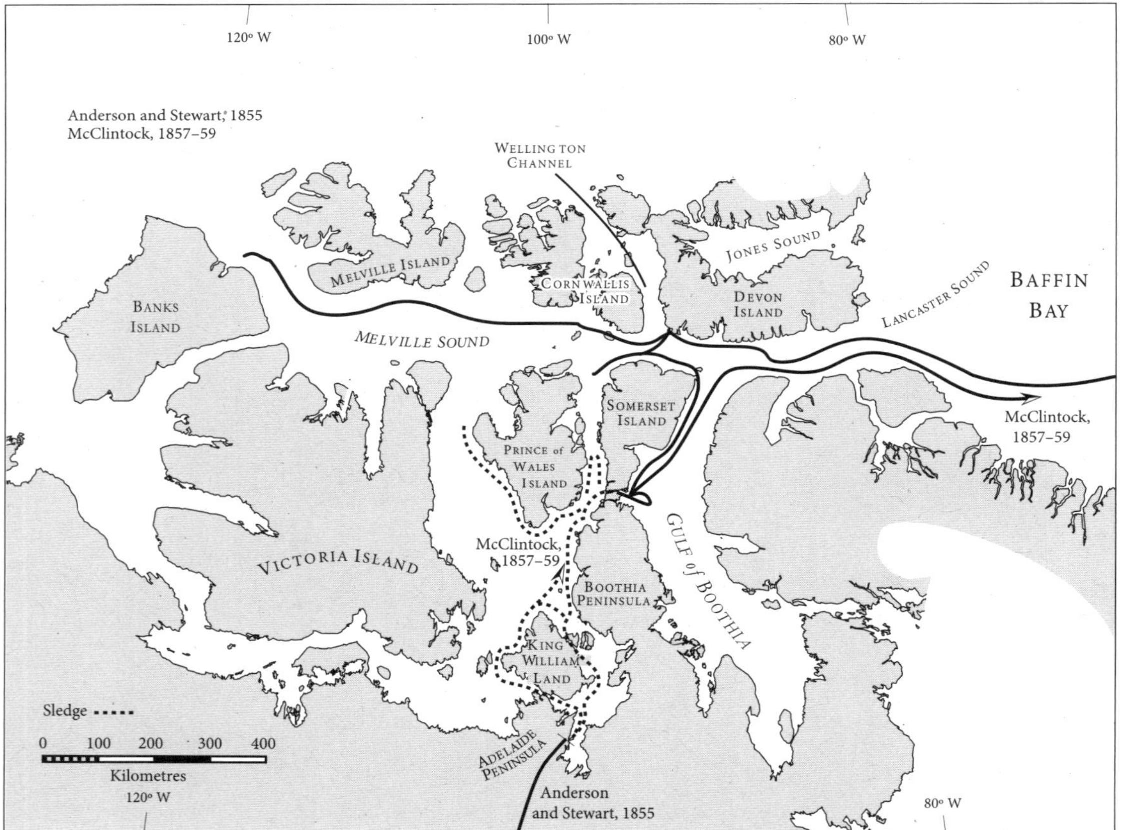

Map 5
Expeditions,
1855–1859

Map by Michael Pacey

As affecting the fate of my absent husband

INTRODUCTION

When Captain Leopold McClintock arrived back in England in 1859 after two years in the Arctic ice and with the secret of Sir John Franklin's fate in hand, he provided the conclusion to what was for many Britons the greatest mystery of the age. McClintock had been in the Arctic looking for survivors or the remains of Sir John Franklin's expedition, whose members had been missing since 1845. His voyage north was the last in an emotional and politically fraught series of government-sponsored, privately funded, and international searches conducted over a twelve-year period. In addition to finding piles of garbage and human remains on his overland journey around King William Island, McClintock and his crew discovered a single piece of paper under a cairn at Victory Point, on the island's north coast, that told of Sir John Franklin's death in 1847 and of the surviving crew members' plan to head south. This "provokingly laconic" document suggested what McClintock was able to confirm as he continued his search: that the men from Franklin's expedition had died while crossing the Northwest Passage they had set out in 1845 to discover.[1]

Back in England, the "Victory Point Record," as the note came to be called, was accepted as the final word on the expedition's fate. McClintock's account of his discovery was published to great acclaim, and the question of Franklin's disappearance was considered resolved. Other than the cryptic and outdated note, written before the majority of Franklin's party had died, McClintock had not found any "record, journal, pocketbook, or memorandum of any description" from the missing expedition.[2] Since that single discovery, no further significant papers from the expedition have come to light. This dearth of first-hand written evidence has contributed to the Victory Point Record's current hallowed status as the single "official text" of the lost Franklin expedition. This single-page document does not deserve such status, however. It is just one of the many disparate threads in the complex narrative tapestry of the expedition, comprising press editorials, Admiralty documents, private letters, and explorers' reports. While all these documents are interesting in their own right, they each gain significance by being part of the larger, cohesive saga of the Franklin expedition. What holds these diverse narrative elements together is the writing of Jane, Lady Franklin.[3]

Lady Franklin, the wife of the famous navigator whose 1845 disappearance into the Arctic ice with 128 crew members became an international incident, did not write books or articles on the subject of her husband's disappearance. Nor did she write fiction for general consumption. She wrote letters, and her letters became the underlying text on which the majority of Victorian and later works on Franklin were constructed. This is not to say that her letters were directly quoted by writers celebrating Franklin or that she contributed any primary material to their projects. Rather, her letters inspired others to persevere with the searches when the Admiralty would willingly have abandoned them; they kept her husband's memory and honour alive in the public imagination; and they provided a heroic template of Sir John Franklin that influenced both the explorers who went in search of him and the writers who later celebrated his adventures. Lady Franklin herself never went north to search for her husband, and all her opinions were based on other people's accounts and reports. But what Victorians learned about the lost Franklin expedition – and thus what we largely know about it today – they learned as a result of her fervour to locate her husband and, if he could not be found alive, to locate his honour in his discovery of the Northwest Passage.[4]

Pierre Berton argues that in strenuously urging her husband's rescue Lady Franklin "added a footnote to history."[5] This collection of her writing is intended to show that Lady Franklin's involvement in the search for her husband was much more significant than that. Her letters show that she was directly responsible for the expedition that discovered Franklin's fate (for which she received, in 1860, the Royal Geographical Society's Founder's Gold Medal – the first woman so honoured), and she almost single-handedly resurrected her husband's reputation from ignominy. Moreover, the searches she so strongly advocated contributed to the opening up of one of the world's most mysterious places. Without her involvement, the Northwest Passage – and her husband's fate – would have remained an elusive mystery for many more years. Lady Franklin's determination that her husband and his crew be found and rescued led her to write directly to public figures whom she might not otherwise have approached so openly. Over the course of the decade-long search, Lady Franklin corresponded with American politicians and entrepreneurs, members of the prestigious Arctic Select Committee, the Lords of the British Admiralty, and British members of Parliament, including two future prime ministers. Combining heart-felt emotion with powerful rhetoric, Lady Franklin penned letters to prominent political figures in England and beyond that were strikingly personal and profoundly universal, appealing to the inherent generosity of humanity through her own private fear and grief. As a collection of private appeals to public figures, as a lone woman's correspondence with a cadre of powerful men, and as a record of a domestic perspective on an international disaster, Lady Franklin's letters provide a unique window into the discovery of the lost Franklin expedition. In contrast to explorers' reports, Admiralty Blue Books, and Victorian press editorials on the subject, Lady Franklin's letters offer us the opportunity to read of the tragedy as it unfolded. Her hopes and fears, her devastating acknowledgment of her husband's death, and her efforts to restore Franklin's honour and shine a light on his heroism are revealed in her forceful and elegant prose. As an advocate for all of the missing sailors, Lady Franklin became an enduring symbol of faith and fidelity. Through her letters, she transformed a tragedy – one that many Britons believed was best left forgotten – into a celebration of British fortitude, faith, and heroism.

When she began her crusade in 1848, Lady Franklin had one main objective: she wanted to rescue her husband. Failing that, she wanted his name to retain the honour she felt it deserved. Sir John Franklin had been a national hero after his first two Arctic expeditions (1819–22, 1825–27) and his command of a frigate in the Mediterranean (1830–33). His reputation, however, had suffered during his much-maligned tenure as lieutenant-governor of Van Diemen's Land (now Tasmania) from 1837 to 1843. Historians universally surmise that his third and final Arctic captaincy was meant to restore his national reputation and perhaps his personal pride as well, but Franklin's disappearance after 1845 threatened his fragile reputation even further. It thus became a priority of Lady Franklin's to identify her husband as a hero under any circumstance.[6] When Franklin went missing, his competence as a leader was called into question, and this doubt, along with her desire for his safe return, was the catalyst that prompted Lady Franklin to action. In order to bring about the rescue of Sir John (or his reputation), Lady Franklin developed and maintained an extensive, as well as delicate, network of relationships: she cultivated a productive acquaintance with the Admiralty (which had the power to supply her with ships, men, and stores, or to refuse her aid altogether); she kept the public interested in her cause; and she vigilantly and vigorously refuted any arguments she felt were detrimental to her chances of success. The only weapon available to her to accomplish these objectives was the letter, and she used it to her advantage throughout her campaign. The letters and documents in this collection reflect all three elements of Lady Franklin's efforts as they trace a clear chronology of her hopes, fears, and determination. They present a picture of a woman who stands firm against a terrible calamity, willing it away through the force of her faith. Read with the hindsight of history, these writings also chronicle the tragic progression of her growing awareness of her husband's death. Lady Franklin's later letters (those written after 1854), fully acknowledging the loss of Sir John, transform her crusade for a living husband into a quest for a heroic one.

One of the challenges Lady Franklin faced in entering the public discussion concerning the search for her husband was her inevitable celebrity. As Franklin's reappearance in England became alarmingly overdue, leading articles in the London *Times* and other daily newspapers increasingly took up the subject of the search for him and his ships. Clearly, the issue of Franklin's disappearance and rescue was a matter of public interest, and people frequently voiced their opinions in the press and to the Lords in Whitehall. Lady Franklin was both a participant in and one of the focuses of the debate. As a regular citizen, she was simply one of many armchair adventurers offering advice and help to a reluctant Admiralty.[7] As the representative of the missing captain, however, Lady Franklin was also a public icon of responsibility. She worked tirelessly to sustain the public's interest, while her continued activities rebuked the Admiralty's indifference.[8] Despite the fact that she did occasionally venture out into public in order to garner support for the search – in the spring of 1848 she visited the Orkneys to hear any tidings of her husband first-hand and to distribute her reward poster to whalers personally – she still claimed to be surprised to see her name in print. In a June 1849 letter to politician and family friend Benjamin Disraeli, for instance, she admitted she was both "thrilled & vexed" to see one of her letters, a copy of which she had recently sent to Disraeli, "ushered into notice" in the *Times* (see document 8 in the collection in this book).[9] To be both thrilled and vexed is to be pulled in two directions, and this brief description reveals Lady Franklin's discomfort with her position. While she hated the idea of being the sole representative of such a public campaign, she recognized that to sustain interest in her cause there was a need for publicity. And support for her cause, at least during the early years of the search, was high. Letters such as the one reprinted in the *Times* that she had been "vexed by" were brought to the public's attention as epistles that would be "properly appreciated in this country, as well as by the officers and men of the missing expedition."[10]

Because of her personal stake in the outcome of the debate about the searches – and perhaps also because of her obvious knowledge about the topic – Lady Franklin was compelled to play a large part in the public conversation about her missing husband. And Lady Franklin's letters certainly fascinated the interested reading public, which duly acknowledged her role as an active participant, not a passive wife, in the debate over her husband's fate. In its 1857 review of a recent letter to Lord Palmerston, for instance, which Lady Franklin herself had published as a pamphlet for general distribution, the *Athenaeum* paid specific homage to her written powers of persuasion, identifying Lady Franklin "in the character of authoress" and celebrating "her extraordinary exertions [which] have been very instrumental in promoting the efforts that have been made to find the lost Expedition."[11]

The public debate on the Franklin search that took place in the print media of the time was not simply a series of back-and-forth letters to the editor between staff writers and random members of the reading public. In what Laurel Brake terms "the cacophony of discourses" of Victorian periodicals, non-fiction essays, editorials, and letters to the editor (such as the one that introduced Lady Franklin's letter to the *Times* readers) played key roles in the intratextual conversation, providing larger relevance and cultural context to specific aspects of the debate.[12] It was precisely the vibrancy of this public conversation – and the innate interest of the topic – that made the question of Franklin's fate so fascinating to the reading public, allowing the story to stay in the public eye as long as it did alongside other issues of national, cultural, and political interest. Critical summaries of the searches, as well as reviews of recent literature on the subject, were printed in dailies such as the *Gazette* and the *Times*, and in weekly, monthly, and quarterly magazines such as the *Athenaeum*, *Fraser's Magazine*, and the *Quarterly Review*.[13] Reviews of the published journals of some of the searchers and the official Blue Books of the Admiralty, as well as news of discoveries, were printed in these magazines according to the interest they generated, appearing next to literary reviews and weekly social news. Lady Franklin's epistolary presence in these publications was unusual for the period and attests to the level of her celebrity. According to Carol T. Christ, under normal circumstances the sphere of these journals was almost entirely a masculine one. Women were not encouraged to inhabit the discursive space of periodicals, which tended to serve as a "locus for an image of masculine heroism."[14] The debate on the quest for Sir John Franklin was itself implicitly a discussion on yet another form of masculine heroism – Arctic exploration itself was a privilege reserved for "the best of British manhood" – and the "great men" of the Royal Geographical Society and the Arctic community were frequent contributors.[15] Although Lady Franklin's participation was thus even more unusual, perhaps one of the reasons why her efforts were so effective was that she did not challenge the apex of masculine honour represented by Arctic navigation – quite the opposite. Her continued belief in "that phalanx of gallant and devoted men in hard conflict with nature" actually smoothed her correspondence with the Admiralty, for it allowed her to sweeten her pointed criticisms with compliments (document 16).[16]

Her continued belief in Britain's institutions, even as they disappointed her, highlights Lady Franklin's awkward social position in the 1840s and 1850s. As Anne McClintock attests for other Englishwomen in the nineteenth century, Lady Franklin was not simply a "hapless [onlooker] of empire."[17] McClintock claims for women a difficult space during the Victorian period of British expansion, showing them as both "acted upon and acting" within the sphere of imperial attitudes and ambitions. Though writing in a previous era (1792), Charlotte Smith offers a description of the frustrating position in which women such as Lady Franklin found themselves: "Women, it is said, have no business with politics – Why not? – Have they no interest in the scenes that are acting around them, in which they have fathers, brothers, husbands, sons, or friends engaged?"[18] Not allowed to contribute corporeally or materially to the business of empire, women were nevertheless participants in it: they reaped its privileges, perpetuated and challenged its attitudes, and were bound by its rules. And despite the apparent "disinterested" nature of Arctic navigation, it was for all intents and purposes an imperial enterprise. On behalf of Britain, explorers laid claim to the land by drawing and naming it on a map, even if it held no commercial appeal. The attitudes of cultural confidence and the sense of intra-European competition that sent Britons up the Nile were the same as those that propelled them north: they relentlessly pursued an intellectual as well as a commercial imperialism.[19]

As a participant in that milieu – and, indeed, as a recipient of many of its benefits – Lady Franklin embodied many of the attitudes that today make us so uncomfortable with the Victorian period. Particularly as reflected in the letters collected here, Lady Franklin's attitudes towards Native people placed her firmly within the same imperial camp as her husband. She believed that innate British superiority inevitably leads to triumph over adversity and that the "Esquimaux" were a race not to be trusted. Her belief, shared by her husband, that "where Esquimaux can live, there also can Englishmen, with their superior intelligence and larger appliances," became a frequent argument in her defence of more searches (document 16). Even though certain expeditions were expressly ordered to "communicate with the various tribes of Esquimaux, in expectation of gleaning some tidings of the discovery ships," Lady Franklin's distrust of Native intelligence and veracity remained – and it certainly didn't help that the information supplied by Inuit hunters tended to cast doubt on her husband's reliability.[20] Lady Franklin's relationship with imperialism and her support of its praxis in the North were far from straightforward, however. In one sense, Lady Franklin was herself a domestic victim of British imperial exploration. Her early biographer, Frances Woodward, records her subject's frequent illnesses after her husband's disappearance. Lady Franklin never saw her husband again after he set sail across the Atlantic under the British ensign in 1845. She spent almost her entire fortune searching for him and was disinherited by her father as a result.[21] She was also arguably an actor on the imperial stage, however. Pierre Berton and Ken McGoogan claim that Lady Franklin was in fact the one who urged her husband to go north in 1845 and orchestrated his election to his final captaincy, and that she never once questioned the value of the enterprise that killed her husband – a question frequently articulated by others in the press.

As both perpetrator and victim, Lady Franklin had to maintain a fine balance of criticism of and support for the Admiralty, since she was all too cognizant of her reliance on the Lords at Whitehall for the continuance of the searches for her husband. She wanted to be involved, but she did not want to act alone. Lady Franklin's letters were thus a curious mix of blame and praise, scoldings and suggestions. On the one hand, were she to criticize too vehemently, she would simply be shunned. The Admiralty's treatment of Richard King offered a chilling example of the power of personality to affect its interest in a cause, and Lady Franklin heeded the lesson. King, a capable explorer whose ideas as to Franklin's whereabouts ultimately proved to be correct, was hindered by his penchant for pointing out the Admiralty's failings and was as a result universally ignored.[22] On the other hand, were Lady Franklin not to criticize enough, the Admiralty would not feel suitably compelled to make good that which she found wanting in their efforts.

Lady Franklin's letters are significant for having successfully pushed against the official grain, but her correspondence is also important as a unique addition to the literature on the search for Franklin. Generally, nineteenth-century Arctic writing came from three main perspectives: that of the official, that of the press, and that of the explorer. Official documents concerned themselves mainly with scientific information. Latitudes, longitudes, geological formations, and daily mean temperatures were recorded in countless charts and tables. Such documents provided useful data, but only those sufficiently trained could use them to conjure up a picture of the Arctic. The press offered the opposite perspective. Filled with journalistic conceits and often scathing opinions on the quest for scientific knowledge, the articles and essays in newspapers and journals tended to turn scientific enterprise into adventurous pantomimes, giving the reading public the opportunity to learn what explorers had for breakfast and what they did in their spare time. The explorers' own journals did much to reconcile these two opposing styles. The opinions they conveyed in their narratives were better informed than those of the papers and magazines, and their scientific observations were often tempered by first-hand experience of the phenomena, though their observations still tended towards the impersonal.[23] Lady Franklin's letters contain elements of all three of these perspectives – they are geographically precise, they are highly dramatic and exquisitely written, and they provide well-informed opinions on one of the period's most interesting topics. But they also offer an important and unique fourth perspective: the domestic one. Of all the unofficial contributors to the Arctic debate, no one was as personal as Lady Franklin. While the press indulged its own penchant for drama, it often undermined the human element of the story. The *Times* awed readers with its horrific descriptions of "the bleached bones of [England's] children" lying in the "dull and lifeless region that guards the axis of the world," but Lady Franklin's often euphemistic descriptions, such as that of "the crushing misery of Dr Rae's tragic story," were much more evocative of the tragedy (document 20). Explorers' journals and accounts in the press often traded on the exotic locale of their subject; Lady Franklin's letters did the opposite, offering instead an emotional and domestic glimpse of those left waiting anxiously behind in England. Indeed, when she claimed to speak for herself and her "fellow sufferers in calamity," she referred not only to the lost men, but also to their family members in England, giving voice to a hitherto silent side of a national disaster and bringing to the Admiralty's attention the domestic consequences of its practices (document 23). Lady Franklin's knowledge of Arctic matters was both broad and minute. According to one admirer, she was "closetted [*sic*] from morning to night … in a private suite of apartments near the Admiralty, day after day, week after week," studying maps, reading sailing instructions, and interviewing Arctic navigators in order to learn all she could about the best methods of relieving her husband.[24] In each of her letters, however, Lady Franklin traded not on the quality of her often lengthy scientific analyses, but on her "language of deep emotion," which underscored her scientific observations and refuted the transformation of her husband's disappearance into an academic question (document 16). In spite of her intimate knowledge of Arctic matters, Lady Franklin steadfastly remained the "afflicted wife" of the missing captain, refusing to depersonalize her arguments and holding on to her emotional connection to the missing expedition as the source of her power and voice.[25]

In her correspondence, Lady Franklin in fact spoke with two voices: that of the Arctic expert and that of the wife. These voices were separate but mutually complementary – the "woman's voice" often framing the Arctic analysis – because they denied the Admiralty any rhetorical room to manoeuvre. Using the voice of the Arctic expert, she called on the Admiralty to perform its official function and rescue her husband, sweetening her argument by alluding to the remainder of the Arctic that lay tantalizingly blank on British maps. Using her own voice as Franklin's wife, she then changed tack to define the search as an emotional and moral imperative. Lady Franklin's two discourses were distinct both in style and rhetoric, together effecting a kind of enclosure for the information she included: no piece of information in her letters was refutable on either scientific or emotional grounds. The sections in her letters devoted to Arctic analysis were clear and straightforward; in them, she analysed the reports of returned navigators and discussed routes of passage for ships into the pack ice and the efficacy of different types of travel. The writing in these sections was academic, and often the only indication that Lady Franklin had more at stake than an answer to a scientific problem was her occasional reference to "my husband Sir John." In the middle of a sustained discussion on routes of travel, however, even these brief articulations of personal connection occasionally slipped into impersonal references to "Sir John Franklin."

For all Lady Franklin's force in the academic portions of her letters, however, the real power of her writing rested in her emotional appeals, as these appeals denied her readers the right to defend themselves on purely academic or scientific grounds. She apologized for her pointed and accurate opinions on the Admiralty's behaviour even while she made them, excusing her behaviour by citing her emotional distress. Elizabeth Harries terms these apologies "conventional female motions" inherent in women's writing of the period.[26] Certainly Lady Franklin must have felt some trepidation addressing the Admiralty Lords with so many and such elaborate requests that would justify her motions of submission, but these "conventional female motions" were in fact part of Lady Franklin's rhetorical strength. The introductions and concluding remarks to these letters could be those of a writer grossly subordinate to her reader, a possibility Lady Franklin acknowledged frequently. After all, she depended heavily on the Admiralty for supplies and manpower for her private expeditions. Apologizing if any "hasty expression, such as [she] ought to have avoided, has escaped [her] pen," however, Lady Franklin would respectfully excuse herself from criticism by summoning up the purpose for her writing, which effectively placed her above rebuke: "I [have] a husband's life at stake for my excuse" (document 16). But while she regularly expressed "doubt whether [she] should have written as [she had] done" in light of her gender as well as her audience (document 16), she never retracted her arguments or criticisms from the Arctic sections of her letters. When she confessed that perhaps "[her] advocacy must be weakened, perhaps even [her] facts suspected" because of her too close connection to the subject, she herself knew, through her own studies of her husband's instructions and her numerous interviews with Arctic experts, that her position – academic or emotional – was irrefutable.[27]

Although Lady Franklin kept up a focused and regular epistolary campaign for the duration of the searches (1848–57), her historical identification as Sir John Franklin's determined wife derives mostly from the earlier half of her writing career. In 1848, as noted above, she travelled to the Orkneys to distribute her own notice of reward to the British whalers and to ask after news of her husband among captains and crews of returned ships, and between 1850 and 1853, she purchased two ships, financing and organizing three private expeditions of her own. In addition to sustaining her own activity, she inspired others to join her cause during this period, writing her most successful letters and reaching even international audiences. American, French, and Canadian adventurers joined the search, inspired by Lady Franklin's appeals.[28] Although North American readers were already eager to follow the path of Franklin through the high Arctic (in 1847, for instance, the northern explorer David Thompson had offered his opinions in the Montreal *Gazette* as to the whereabouts of the missing expedition), it was Lady Franklin's petition on her husband's behalf that prompted the American government in 1850 to outfit its own expedition to go in search of the missing Britons (document 4). In her April 1849 letter to US president Zachary Taylor, Lady Franklin appealed to the president's sense of diplomacy and international politics, but she also appealed to his humanity, highlighting her own personal stake in the expedition in a way that neither the Admiralty nor another, official correspondent would have been able to do. Lady Franklin cited the very personal "intense anxieties of a wife" as the impetus for her correspondence. President Taylor may have already had a predilection towards altruism, but Lady Franklin's personal outline of her dilemma and her simple appeal to his humanity struck a particular chord in him. When the letter was published in the American papers to garner national support for the cause, it caught the philanthropic eye of another prominent American: Henry Grinnell. A shipping magnate from New York, Grinnell was so moved by Lady Franklin's appeal that he gave a substantial portion of his private funds to the American search parties and began a private correspondence with her that reached beyond the search for her husband.[29] Out of Lady Franklin's first humble letter to Taylor grew years of near-unconditional support. Between 1850 and 1856, two American expeditions were conducted with the support of Grinnell and the American government, and in 1856 the government sanctioned $40,000 (US) of public money to re-equip a British brig abandoned in the ice in 1854 during a search for Franklin and return it to England in "token of the deep interest felt in the United States for the service in which [that] ship was engaged."[30]

Lady Franklin's correspondence was equally inspiring at home. Her epistles – "worthy of a sacred cause" – were frequently quoted and referred to in Parliament, printed in newspapers, and used to support arguments for the continuation of the searches.[31] While privately she may have balked at her own visibility, she understood that her celebrity and identity as "a national monument of fidelity" furthered her cause.[32] As alluded to above, her persistent epistolary presence at Whitehall and her public eloquence in urging her husband's rescue naturally made her an advocate for the family members of the missing sailors as well. Just as her husband had led his men into the Arctic in search of the Northwest Passage, so Lady Franklin, his representative at home, led the relatives of the missing crews to the desks of the Admiralty Lords. Though some might have argued at the time that advancing on the Admiralty in Whitehall was more dangerous than battling icebergs in the North, Lady Franklin was impressively equipped for the task. A January 1850 letter to Lady Franklin from Dr Robert Goodsir, brother of Assistant Surgeon Harry Goodsir, who was aboard the missing *Erebus* (Franklin's own ship), was typical both in its straightforward request for Lady Franklin's support and in its implicit confidence in her power to help. After explaining that he had heard nothing yet from the Admiralty regarding its plans for a search, Goodsir appealed to Lady Franklin with the expectation of quicker action: "I hope I am not asking too much in requesting your good offices in obtaining me an appointment of any kind in any of the

Expeditions that may be sent out during the next season."[33] On several occasions, Lady Franklin wrote to the Admiralty expressly to recommend such appointments to the searching crews, and on 23 January 1850 she wrote on behalf of Goodsir (document 11).[34] Goodsir's faith in Lady Franklin's ability to obtain him a position was a compliment to her powers of persuasion; that Goodsir sailed aboard the *Lady Franklin* later in 1850 (commanded by another of Lady Franklin's recommendations, Captain William Penny, mentioned in the same letter), was proof of her tenacity. Lady Franklin's determination to do her part in the rescue effort was a constant inspiration to others, and her role as an exemplar did not go unappreciated. When Robert Goodsir returned to England and published his narrative of the search expedition, he praised Lady Franklin for providing him the opportunity to do his part, drawing attention to her role as Admiralty whip: "Much, very much, gratitude is due from the friends and relatives of all the missing voyagers to the noble-hearted and never-tiring Lady Franklin, for stirring up the energies of some, stimulating the forgetful, and shaming the careless, into renewed efforts on behalf of our fellow-countrymen."[35] At that time, there was little doubt in the public's mind who was meant by the "careless" and "forgetful"; much of the Admiralty's behaviour towards Lady Franklin had been "unhappily in no way calculated to add to [its] renown."[36]

Another of the reasons Lady Franklin was such a formidable opponent for the Admiralty was that she was not alone in her deep concern for the missing Franklin expedition. Even Britons with no familial stake in the outcome of the expedition followed the ships – and then the search for them – with avid interest. Indeed, for a time the entire nation seemed to be absorbed with the expedition – with the government's encouragement. The search for the Northwest Passage during this period could almost be called a traditional British activity, propelled as it was by sustained Admiralty interest throughout the preceding three hundred years. In the second half of the eighteenth century, successive private expeditions vied for the £20,000 reward for its discovery promulgated by the Admiralty in 1745, and the explorers brought back from the North tales of game, bounty, and settlement opportunity. Though it was eventually determined that the passage would never be practicable as a trade route, by the time explorers and speculators admitted this, it was virtually a moot point, the interest having already shifted from commercial to intellectual conquest. In 1775 the Admiralty amended its reward and signalled a significant change in the way it viewed Arctic exploration: it now offered £5,000 to the person who located the North Pole, indicating a growing interest in geographical discovery. Thus, after disappointment over its unusable trade route, the Arctic was resurrected as an area of academic importance, and the desire to fill in the Arctic map with British names became a viable enterprise for a navy trained in war and political conquest.[37]

Arctic exploration had ceased during Britain's war with France (1798–1815), but following the Napoleonic Wars the possibility of Admiralty-supported scientific conquest became an especially attractive employment option for junior officers. With an inflated population of officers with little chance of promotion,[38] the Royal Navy reinstated the unclaimed £20,000 reward, opened the contest to government-supported expeditions, outfitted its own ships and men for the task, and reintroduced the search for the Northwest Passage as a national adventure. The passage provided necessary employment and distraction to the idle officers and sailors and, as Fergus Fleming notes, served a practical purpose for the officers themselves, who went north "for the chance of employment" and the secret hope that "it might lead to promotion."[39] For the most part, the renewed focus on the passage provided the Admiralty and the officers with a mutually satisfactory arrangement: the Admiralty could legitimately justify enormous expenditures by claiming new territory, while officers received much-appreciated work and pay.[40] Arctic exploration expanded the Admiralty's intellectual as well as territorial horizons. The men sent north to explore new coastlines were transformed from explorer-adventurers into explorer-observers, and the Admiralty moved into the expanding field of science through the observation and recording of natural phenomena. In 1819 Captain William Edward Parry and his crew received £5,000 from the British government for "being the first ship to cross the 110th parallel of longitude," and thus began the Admiralty's own race against itself for the geographical conquest of the North.[41] Nevertheless, while many of the officers quickly warmed to their new career paths, the enterprise was not without risk. Expeditions were sent forth with the newest scientific equipment, anti-scorbutics, and innovative contraptions for overland travel, but little could be done during this era of wood and sail to battle the Arctic's extreme weather. Frequent storms could capsize vessels, and the deadly pack ice threatened to crush entire ships the instant they were stuck. Even if they were not crushed, many ships were pinned in for years on end. Such dangers were part of the Arctic's allure, for they offered appropriate challenges to leadership, skill, and cultural values in the same way that battles did, but without the imperative for bloodshed. In going north, Britons not only packed their compasses and standard-issue woollens; they also took with them their growing belief in the Englishman's ability to survive anywhere and to triumph over any adversity through faith, scientific objectivity, and superior spirit.

The expenditure for Arctic exploration was immense and its tangible benefits unclear, but at least in the early years of the nineteenth century (1818–31), northern discovery was a fairly profitable venture for the government – not in terms of revenue, but in terms of public interest and support. (There were, in fact, some economic benefits to the enterprise – if relatively small. The Admiralty granted the publisher John Murray the exclusive rights to publish captains' diaries [such as those Franklin kept during his first two expeditions] upon their return, and no doubt the Admiralty was a partial recipient of the royalties the sales generated.)[42] Beyond the issue of monetary gain, however, Arctic exploration was valuable to governments as a successful public performance, a needed focus for the navy (where otherwise there was none) and a distraction from the petty politics of peacetime. As Charles Dickens wrote in 1859 of the previous forty years of Britain's Arctic adventures, "Few subjects of national interest have sunk deeper into the national mind."[43] Indeed, Arctic exploration was in itself an expression of English physical vigour and intellectual curiosity. In his best-selling journal from his first captaincy (1819–22), John Franklin recorded his admiration for the English seaman, describing him as someone who would enter "upon any enterprise, however hazardous, without enquiring, or desiring to know, where he is going, or what he is going about."[44] In terms of Arctic exploration, this blind enthusiasm was particularly convenient, for neither the captains of the ships nor the Admiralty actually knew precisely where the expeditions were headed, and there was no practical reason to wonder what one was "going about" in the first place.

Arctic exploration was both a political and a personal point of interest for Britons because it reflected characteristics they found attractive in themselves: the enterprise was relatively objective (the interest in capturing and abducting Native people or establishing settlements had waned by the late eighteenth century); it was logical (it was dominated by scientific principles of observation and record-keeping); and it was noble (it was not done overtly for monetary gain). Arctic exploration was a "clean" enterprise. It claimed one of the last remaining "blank" spaces of the world's map for Britain, and as Sir John Barrow put it in 1818, it was a project "worthy of a great, a prosperous and an enlightened nation."[45] The explorer in charge of such a noble undertaking took on the sheen of heroism, joining the cadre of what Lisa Bloom terms the "secular hagiography" of Arctic heroes.[46] Just as the quest for the magnetic true north seemed devoid of political agenda, the polar region itself drew "all that is right-headed and right-hearted from the sailor" who searched for it.[47] Like the enterprise itself, the explorer and his ships represented the best of Britishness: the Arctic explorer was clean, Christian, and cut a somewhat more genteel figure than the explorer of Africa; his ships were self-contained (a little piece of Britain in the open ocean); and wholesome entertainment for the crew was the official rule.[48] The Arctic was the "one unspotted place" in the world where men could adventure without the attendant politics – Arctic navigators battled icebergs and storms, not armies and governments.[49] Frozen ships and men on half-rations embodied what many considered an Arctic idyll. Despite the occasional harrowing encounter with an iceberg, the danger of being frozen in, and the phantom of scurvy, the traditional wartime strictness of life aboard ship gave way to printing presses, plays, and poetry.[50] In the official published narratives, as well as in the editorial speculations on expeditions not yet returned, audiences back home could read what the explorers put in their Christmas pudding, could see what they wore to keep warm, and could appreciate such artistic forays as hand-drawn renderings of icebergs shaped like cathedrals. If Britons took a personal interest in Arctic explorers, it was because they had become familiar with what it meant to be an Arctic explorer: the details of daily existence had been made visible, and what they saw they liked. As Henry Morley wrote in 1853, and as the general public agreed, "The history of Arctic enterprise is stainless as the Arctic snows, clean to the core as an ice mountain."[51] Certainly men died while on expeditions north, whole ships were sometimes lost to the pack ice, and some crews were left for dead; but by simply returning to England whatever way they could, the intrepid explorers themselves showed that it was possible to overcome terrible odds and to triumph with vigour over a unique brand of adversity.

Rudy Wiebe takes a more jaundiced view of the challenges, terrors, and tragedies of nineteenth-century British Arctic exploration, arguing that the "historical record implies that the greater the numbers of their fellow heroes [that the British officers took] with them in death, the greater the so-called heroism and of course the proportionately larger the resulting memorials."[52] Wiebe was referring to Sir John Franklin's first expedition (1819–22), which lost eleven of its twenty men to starvation and murder, when he wrote this admittedly "deeply cynical" statement, but as he says, the historical record does seem to bear out his assessment.[53] Indeed, on one level, Franklin's Arctic experience does support Wiebe's statement: Franklin remains the most famous of the nineteenth-century British Arctic explorers instead of the most infamous, even with the dubious honour of having lost the most men on an expedition (129, including himself). In the context of Lady Franklin's letters, however, it is possible to look at the celebrated "heroism" and fame of the perished men as less a consequence of their misguided ambition and more a product of the love and admiration the living continued to feel for them. Dead men could argue nothing for themselves – a fact Lady Franklin gently pointed out in one of her letters to the Admiralty (document 22) – and Lady Franklin and her supporters certainly considered it the duty of the loved ones left behind to recover the honour of the missing.[54]

The Admiralty seemed all too happy to cut Franklin adrift in its sea of forgotten adventurers. At the end of March 1854, it announced that he was considered to have died in Her Majesty's Service, and it granted prize money for geographical discovery to explorers who had ostensibly been looking for Franklin's ships, thereby gradually depleting the available funds and incentive for further searches. As search expeditions continued to return empty-handed, the press came largely to support this official position. Though increasingly isolated amid growing public apathy and government disinterest, Lady Franklin continued to insist on additional searches, arguing that her husband was the hero she believed him to be and insisting on the opportunity to prove it. Indeed, through their continued advocacy, Lady Franklin and her supporters were solely responsible for restoring Sir John Franklin's reputation (and that of his crews). With no recovered narrative of Sir John's expedition set down in a sailor's own hand, with their own letters and own rescue expeditions Lady Franklin and her supporters wrote Sir John and his crew members back into a history that was prepared to forget them.

There was a variety of reasons why Franklin's heroism was particularly hard-won. Of all the expeditions sent out by the British government of the period, Sir John's was the one considered most likely to succeed: it had roughly thirty years of relatively unbroken experience behind it; the majority of the external coastline of the eastern Arctic had already been mapped; its ships were laden with the latest in explorative technology (they were equipped with giant screw propellers powered by steam engines designed to push the ships through the floating pack ice, and their larders were stocked with canned goods); and numerous explorers had already shown that expeditions could get lost for years, "fail" in their original goal, run afoul of weather and climate, yet still come home with an impressive list of discoveries and some ripping good yarns.[55] That there was such widespread confidence that Franklin would succeed made the possibility of his failure flagrantly ridiculous – it assumed a level of incompetence that could not be entertained. (As Berton acknowledges, had there been even a suspicion in 1845 that the Franklin expedition would meet with trouble, it would not have been sent at all, since it was already the most costly Arctic expedition to date.)[56] And, indeed, Franklin himself had thought his success was in hand the moment he set sail. As members of the Admiralty would hear from whalers who had visited Franklin's two ships, the *Erebus* and *Terror*, before they entered the eastern pack ice in 1845, Franklin had allegedly urged his officers and men "to inform their friends that they could not expect to hear from them after they entered the ice until the month of October, 1847, – when he trusted they would be successful."[57] As Franklin's expected date of return came and went in 1847, Sir James Clark Ross, the most prominent Arctic explorer of the day and the Admiralty's original choice for captain of the missing expedition, remained unconcerned about their continued absence, arguing that, as their provisions could easily "extend to four years without any serious inconvenience," there was not "the smallest reason of apprehension or anxiety for the safety and success of the expedition under the command of Sir John Franklin."[58]

There were, however, unfortunate extenuating circumstances that neither the members of the expedition nor the Admiralty had considered, and that ultimately led to disaster. Specifically, the very technological innovations meant to speed the explorers on their way to fame and fortune quickly became hindrances, leading even the men searching for them to believe the worst of their fate. The propellers, powered by converted railway steam engines and designed to push the ships through the ice, were worse than useless, rusting rapidly, adding unnecessary weight to the ships, and using too much fuel to be practical. Of greater significance to the fate of the expedition was a problem with the food on board. The *Erebus* and *Terror* carried 8,000 tin canisters of food, a three-year supply. The meat and vegetables in the tins comprised the bulk of provisions for the two ships and were to provide the variety and nutrients necessary for the men's prolonged stay in the North. However, in 1850 an unusually large cairn of emptied tins was discovered on Beechey Island at the site of the missing expedition's first wintering quarters, leading some searchers to believe that the food had spoiled. When searcher Sherard Osborn noticed the label of the manufacturer, he lost hope in the missing men's survival, knowing as he did that the supplier had sold a later search expedition a batch of tainted food: "Sad it is to record it, but nearly all their preserved meats were those of the miscreant 'Goldner.' This alone would lead to the disastrous fate of the expedition, by so crippling their resources by condemning them to a diet of salt meats."[59] Complicating this discouraging discovery was the hopeful but tragically erroneous report of one of the whaling captains who had encountered Franklin. Apparently, Franklin had told him that he was confident his provisions could "spin out seven years."[60] This testimony regarding Franklin's own estimation of his food resources was a factor in the decision of when to begin looking for the lost expedition.

These conflicting reports and opinions did little to encourage search efforts, and the public as well as the Admiralty continued to believe that the Franklin expedition was merely following in the footsteps of its predecessors – the longer it was gone, the more spectacular its discoveries would be. Nevertheless, under pressure from Lady Franklin and others, the Admiralty launched fourteen expeditions from 1847 to 1854, the largest comprising five ships and some remaining "in the fastnesses of the north" for up to five years (document 16). Supplementing the government expeditions were numerous private searches, equipped through private means, the Hudson's Bay Company, and public subscription. Lady Franklin herself financed four expeditions. In spite of the occasional crowding of British search ships in the eastern Arctic, however – different expeditions even sometimes overwintered together – the territory covered was insufficient. The first expeditions that set out in search of Franklin (in 1847, 1848, and 1849) were cursory; a number of summer runs to the whaling harbours and a brief tour of the outer eastern coastline were considered adequate in the hunt for an expedition that few considered lost even after its allotted three years had elapsed. The expeditions in the early 1850s were hardly more effective. The captains of these were pulled in multiple directions as they headed north – the cachet of finding Franklin was matched by that of reaching the North Pole, and the Northwest Passage was still a compelling lure, in spite of the obvious dangers. The continued if erroneous belief in an open polar sea in the extreme north also hindered the searches, since captain after captain pushed north, convinced the polar sea was the passage to both Franklin and fame.[61] That Franklin's original instructions were to sail in a southwesterly direction seemed irrelevant under the weight of this belief. Certainly there were officers who would happily have embarked on the search for nothing in order to save their fellow adventurers, but those men for the most part travelled under Lady Franklin's flag, not the Admiralty's.[62]

In 1854 three events occurred that threatened to defeat Lady Franklin and to wipe her husband from the historical and heroic record: the Admiralty pronounced her husband as having "died in Her Majesty's Service" before the results of the most recent expedition were known; the Crimean War began; and the Hudson's Bay Company explorer Dr John Rae purchased from a group of travelling Inuit hunters a number of relics from the lost expedition, receiving along with the small objects news of the crew members' whereabouts. By 1854, the frequency of government-sponsored searches was already declining. Franklin and his men had been gone for nine years, and with few exceptions the Admiralty and public alike were ready to mourn the missing sailors and turn to other things. The 1850–52 period had been the prime years of the search – nineteen ships had gone out under British as well as American flags, financed by public as well as private donors, and the Admiralty, Parliament, and the press had consistently registered respectful concern for the missing explorers. However, as political tensions with Russia rose, the national interest shifted from exploration to politics. Readying for war, the Admiralty could no longer afford to fund and outfit search parties for men now completely unconnected to current events, and for the officers and men still in Britain, the Crimea offered a more direct path to heroism. Still, 1852 saw the outfitting of the largest search expedition the Admiralty had yet put together: five ships under the command of Sir Edward Belcher were readied for four years in the Arctic. In January of 1854, however, a full two years before this expedition was expected back, the Admiralty announced in the *Gazette* that it would strike the missing sailors' names off the registry list if news was not heard of them by the end of the upcoming March. When offered a widow's pension, Lady Franklin broke Victorian protocol and changed "the deep mourning she had been wearing for years [because of Sir John's absence] for bright colours of green & pink," protesting the abandonment of the searches and displaying her dauntless faith in her husband's continued existence.[63] Her sartorial strategy, however, had no effect. In late March 1854 – four days, Berton records, before Franklin was officially declared dead[64] – the Admiralty recalled its ships to help in the Crimean war effort, and the money for exploration and searches evaporated. Belcher's expedition, the only one still searching for Franklin, came home an unmitigated failure: four of the five ships had been abandoned to the ice, and no new discoveries – geographical or otherwise – were made.[65] But most distressing, perhaps, was Dr John Rae's story, coming as it did from an individual not employed in the search but a known expert in northern exploration, and seeming to offer definitive evidence of the missing men's fate.

In October 1854, Dr Rae arrived in London from a surveying expedition in the Arctic with news of Franklin and his men. Rae's private report to the Admiralty was printed in the *Times*, and it was there that England first read of the apparent fate of Sir John Franklin:

> In the spring, four winters past (spring, 1850), a party of 'white men,' amounting to about 40, were seen travelling southward over the ice and dragging a boat with them by some Esquimaux, who were killing seals near the north shore of King William's

Land, which is a large island ... At a later date the same season, ... the bodies of some 30 persons were discovered on the continent, and five on an island near it ... From the mutilated state of many of the corpses and the contents of the kettles, it is evident that our wretched countrymen had been driven to the last resource – cannibalism – as a means of prolonging existence.[66]

Rae admitted in his report that "[n]one of the Esquimaux with whom [he] conversed had seen the 'whites,'" nor had he himself visited the site of the tragedy, but he offered "no apology for taking the liberty of addressing" the Admiralty or for his "meagre" tidings. Meagre as the length and breadth of the news may have been, it was the final nail in the coffin of the Admiralty's search for Sir John and his crew. The Admiralty had no interest in proving incontrovertibly that the missing expedition had in fact succumbed to "the last resource," and the best way to avoid doing so was to abandon all thoughts of any future search. Rae's news of cannibalism made the Franklin expedition more than a national tragedy; it was also suddenly a source of national shame. Lady Franklin thus had to do battle on two fronts: she was determined to refute the stigma of her husband's alleged fate, and she needed to find alternative means of support, given the Admiralty's decision to withdraw from the enterprise.

In order to clear her husband's name, Lady Franklin needed monetary and human resources, but the Admiralty refused her both (though kept her waiting for an answer for nearly three years). Her family hardly gave her more support, her stepdaughter and her son-in-law taking their anxiety over Lady Franklin's spending habits to the papers and, in 1851, her own father disinheriting her from his estate.[67] Lady Franklin was not completely without allies, however, either private or public. Franklin's niece Sophia Cracroft was Lady Franklin's constant companion, often acting as secretary, nurse, and confidante; and Charles Dickens and his cadre of *Household Words* writers, including the aspiring novelist Wilkie Collins, questioned Rae's character and cast doubt on the reliability of his report, arguing on Franklin's behalf that "the better educated the man, the better disciplined the habits, the more reflective and religious the tone of thought, the more gigantically improbable the 'last resource' becomes."[68] Dickens's attack was two-pronged: he defended Franklin's innate honour as a British gentleman, making general statements of nationalist pride, and criticized Rae for not having seen with his own English eyes a scene that could only have been distorted in the "incoherent Esquimaux story" Rae had received.[69] Dickens left no room for interpretation in his assessment of the Inuit and the weight he gave their tale: "We believe every savage to be in his heart covetous, treacherous, and cruel."[70] Regardless of what was said about Sir John Franklin, Dickens's point was that only an Englishman could tell an Englishman's tale.[71] In *The Frozen Deep*, a play about a missing Arctic expedition (first performed at Dickens's home as his annual Christmas pageant in 1856 and then by request for Queen Victoria), Dickens, collaborating with Wilkie Collins, transformed Rae's story into a romantic allegory. It became a tale of murder and revenge over a failed love affair – there was no reference to cannibalism or to Inuit involvement, and the Rae character was turned into a suspicious, power-hungry nursemaid who predicted the expedition's doom in her effort to ruin the happiness of the delicate heroine.[72]

John Rae defended himself and his Inuit informants admirably in the press – he was obliged to do so on more than one occasion – and was awarded £10,000 for his discovery. As a consequence, however, he lost favour, permanently, with the Arctic community. Lady Franklin and her niece Sophia Cracroft accused him of stealing artifacts from the missing expedition, and his testimony, C. Stuart Houston suspects, might have cost him a knighthood.[73] But regardless of the public's outrage over the cannibalism tale, the Admiralty, having already announced Franklin's death in the *Gazette* the previous March, refused to mount a final expedition to search for relics or survivors. Perhaps the most frustrating element for Lady Franklin in John Rae's report was its combination of social horror and geographic specificity: finding the men might confirm the truth of cannibalism, yet the alleged location of the bodies argued that the dead explorers deserved commendation – they were, according to Sir John Franklin's original instructions, right where they should have been. As Frances Woodward points out, the story did provide some relief: "[A]t least Rae had put one fact beyond a doubt: Sir John Franklin had carried out his instructions."[74] Lady Franklin was in the difficult position of having to refute one half of Rae's tale while supporting the other.

In 1855 a new enemy entered the fray. The Arctic Select Committee met to discuss Captain Robert McClure's claim to the reward for the discovery of the Northwest Passage. In 1853 McClure had walked to safety across a frozen passage with his remaining crews after having been locked in the western Arctic for three years in a fruitless search for Franklin. Nearing starvation and crippled by scurvy, McClure and his men were rescued by another Franklin search expedition that had entered the Arctic from the east, thus becoming the first Europeans to enter and exit the Arctic from opposite ends. News of McClure's discovery had reached England late in 1853 – McClure had ostensibly been looking for Franklin through the western Arctic when he discovered the passage. While this news was disappointing enough to Lady Franklin on its own, it became more so when, only a few weeks later, Franklin's death was announced in the papers. In her 24 February 1854 letter to the Admiralty (document 16), Lady Franklin articulated her fears that "the solution of the geographical problem appears to have sealed the doom of [her] unfortunate husband and his brave associates." Lady Franklin objected to the adjudication of the reward because she feared that it would eliminate her husband from any right to fame or heroism, even in death. As unfortunate as this timing was, most people were willing to accept McClure's claim; after all, there was as yet no incontrovertible proof that Franklin had accomplished the mission earlier. Henry Morley, a writer for Dickens's *Household Words*, recorded (as many writers did) that McClure was incontestably the first discoverer of the "long-sought north-west passage."[75] Lady Franklin, however, took the proof of McClure's previously undiscovered and unsuspected passage as added incentive to outfit a last expedition north.

Thus, when the Arctic Select Committee met in 1855 to adjudicate McClure's award for his discovery, Lady Franklin contributed a forceful argument in favour of her husband being given credit for "the first discovery of a navigable passage" (document 21). The strength of Lady Franklin's argument lay in her comparison of the routes taken by her husband and McClure, sailing ten years apart through what should have been the same piece of water. While she insisted that it was not her wish "to attempt to question the claims of Captain McClure," she pointed out that her husband was "specially warned … not to seek [the passage] in the quarter" where McClure had made his discovery, implying that Franklin's path was the more valuable (as it was the first chosen by the Admiralty) and, more disturbingly, that McClure had acted against official orders, had not in fact been searching for Franklin, and was therefore not legitimately entitled to the honour (document 21).[76] Her assessment of McClure's accomplishment in comparison to that of her husband was also an indictment against the Admiralty, which had saved McClure but failed her husband, thus giving fame to McClure by default and shame to her husband. She correctly pointed out that McClure had succeeded only by the "timely assistance of the brave men who were in search of him," implying that Franklin failed only because no one arrived in time to return the triumphant explorers to safety (document 21).[77]

Given that the Admiralty had already expressed its intention to lay the Franklin search to rest – it had been a year since Sir John and his crews were considered as having perished in Her Majesty's service – the attention that the Arctic Select Committee gave to Lady Franklin's appeal was remarkable. Her letter was published in the appendix to the report as part of the official documentation of the committee, and in its conclusions, the committee responded directly to her concerns: "It is scarcely necessary to say that neither in their investigations nor in this report have [the] committee in the slightest degree interfered with anything which it is the object of [Lady Franklin's] noble and touching appeal to preserve. [But the] committee cannot determine the extent to which it may yet be found that Sir John Franklin and his companions had carried those discoveries to which they sacrificed their lives."[78] Although ultimately the committee ruled in McClure's favour (it awarded him £5,000 for his discovery and "as a personal tribute to his energy, bravery, and skill"), its members felt the need to respond to and respect Lady Franklin's argument.[79] What the committee's response to Lady Franklin reveals was her formidable presence within the Whitehall offices – clearly, the Arctic Select Committee and the Admiralty in general felt compelled to justify their actions and decisions at every turn. Its response is more indicative of the Admiralty's unwillingness to engage more men, material, and expense to challenge McClure's right to the award than of any desire to disagree with Lady Franklin herself.

Lady Franklin continued to write to the government, urging that a final search be made of the area delineated by Rae, but her appeals were unsuccessful and, over a long period, went unanswered. Indeed, even before Rae's 1854 report, the Admiralty had begun to refuse Lady Franklin the comprehensive aid for which she frequently applied. In a 31 January 1853 letter to the secretary of the Admiralty, for instance, she wrote of the "painful necessity of [her] endeavouring to supply in some measure ... [the] deficiency in the Govt measures" (document 13). After 1854 her exertions became increasingly unsuccessful. In a letter dated 11 July 1856 (document 23), Lady Franklin alluded to "not [having] been honoured with any reply" to an earlier letter, while her letter of 2 December 1856 to Prime Minister Lord Palmerston (document 24) regarding her final search for her husband went unanswered until the following April.[80] Undeterred by an unsupportive bureaucracy, in 1856 Lady Franklin announced her intention to launch one last expedition in search of relics and any survivors, having been "impelled to this course of action by a solemn sense of duty" (document 26). Getting nowhere with the Admiralty, who refused her the use of a ship the American government had expressly outfitted for a search for her husband, Lady Franklin settled on a luxury schooner, the *Fox*, refitted it for Arctic use, and, through public subscription, financed what would become a two-year expedition into the heart of the Arctic. She readied her modest expedition under the public eye, and as public pressure mounted, the Admiralty relented slightly, giving Royal Naval captain Leopold McClintock the required leave to head the voyage under Lady Franklin's orders and donating spare stores to the ship and crews.[81]

While presidents, Arctic experts, and members of the Admiralty often responded to her appeals (as Goodsir so eloquently stated in 1851), her dedication to the search for her husband and her expression of duty also came under criticism. As the search wore on, ships and men went missing, and war broke out in the Crimea, the Admiralty developed a sullen reticence and some of the press lost their patience. For example, in June 1854, after reviewing the most recent round of published Arctic documents, *Fraser's Magazine* defended the Admiralty for abandoning the Franklin search and criticized the letter Lady Franklin had written in protest in February of that year (published by the House of Commons) as "bordering on the harsh and unjust."[82] Even though the magazine excused the "unjust" language of the letter, given the character of the "exemplary and afflicted wife" who wrote it, it expressed its support for the Admiralty, noting that it had "a duty to perform" for the nation (i.e., in the Crimean War) and giving this priority over its duty to one small party of lost men. Saying nothing, unsurprisingly, of the resources and lives then being poured into the Crimean War, *Fraser's Magazine* concluded its criticism of Lady Franklin by stating that "any further encouragement from that quarter might involve a continuing sacrifice of the lives of brave men." In another vein, the *Times* offered a distinctly ungenerous interpretation of Lady Franklin's decision to outfit the *Fox*. In more than one editorial in early 1857 (when Lady Franklin was readying her final expedition), the *Times* referred both to Lady Franklin and to her efforts as "desperate," sneering at her penchant to sink "the last relics of her little fortune … under the Polar ice" with no expectation of success. Rather than supporting Lady Franklin's private efforts – indeed, it "entirely approved" of the Admiralty's decision to end public searches – the newspaper dismissed her last search as "Quixotic" and freely admitted that it could not "with any reasonable grace advocate the propriety or policy even of this limited attempt."[83] Far from supporting her private enterprise, the *Times* went so far as to disagree with her final method of search (by sea) and relegated the announcement of a public subscription for the outfitting to its advertising pages. But while some of the press would have the public believe she remained unsupported in her claims to duty, Lady Franklin did not stand alone in her convictions. Just as she was the advocate for the missing sailors' relatives during the early searches, in her final expedition she had the support of the "high authorities" of Arctic experts and the "hearty sympathy" of much of the public. Many prominent public figures contributed to the financing of this final expedition, inspired no longer by Franklin himself and his missing companions, but by Lady Franklin's sense of responsibility for the missing men (document 24).[84]

Indeed, Lady Franklin's last effort did galvanize the public once more, despite the *Times*'s inability to support her decision. In choosing his comparatively tiny crew of twenty-four, McClintock was overwhelmed with the "innumerable" applications to join the expedition, in particular those containing "offers to 'serve in any capacity' which poured in from all parts of the country, from people of all classes, many of whom had never seen the sea."[85] Arctic stalwarts such as Sir Roderick Murchison (president of the Royal Geographical Society), Sir Francis Beaufort, and Sir Edward Sabine were vocal and visible supporters of the project, and the general subscription yielded nearly £3,000, with members of the sympathetic public donating amounts from 2*s.* 6*d.* to £500. McClintock finally sailed at the end of June 1857, setting out from a harbour crowded with well-wishers, and if the public lauded the sailors as heroes, their pride and affection for Lady Franklin were even stronger. McClintock tried in vain to "repress the enthusiasm of [his] crew" in order that he might spare his patroness an awkward public display before the ship set sail, but three "prolonged, hearty cheers" broke out as the ship cast off, in homage to the woman who had brought the expedition together.[86]

McClintock's safe return to England in late 1859 was the first in a series of final triumphs for Lady Franklin. Not only did McClintock find the final resting places of many of Franklin's men and discover definitively the date of her husband's death (which had occurred before the expedition disintegrated), but he also restored Sir John Franklin to the history books. In direct accordance with what Lady Franklin had argued all along, McClintock affirmed in his concluding remarks on the expedition that the men who "perished in the path of duty" had persevered "until they had achieved the grand object of their voyage – the *Discovery of the North-West Passage*."[87] McClintock's narrative provided the necessary incentive to restore Franklin to the heroic cadre of Arctic navigators, and the British press did its best to make up for lost time, accepting McClintock's heroic interpretation of the expedition's last days with relief and enthusiasm. The precise descriptions of McClintock's discoveries allowed the public once again to imagine and celebrate the missing men; their deaths were no longer sources of shame, but rather formed the final heroic chapter in England's "modern Odyssey."[88] *Fraser's Magazine,* for one, celebrated the sequence of events in the expedition's final attempt to reach the mainland: "Reduced by scurvy to a state of infantine weakness, we see them in their attempt to reach the [Back] River abandoning every article that could be dispensed with, and finally dropping as they struggle to escape from the barren shores of King William Land."[89] Moreover, it was not merely the tragedy of deaths of the crew members that drew readers to the story, for in the same editorial *Fraser's Magazine* transformed that final march into one of triumph: "[M]ay we not hope that their last hours were cheered by the consciousness that they had done their duty, and helped in accomplishing the great work of discovering a North-West Passage, and that they would always be remembered as having triumphed over difficulties which baffled ardent spirits during three centuries?" This triumph was exactly what Lady Franklin had been demanding since 1848, and the press allowed her the full enjoyment of her success. Periodicals that had previously expressed ambivalence over the continuation of the searches bowed to the incontrovertible evidence laid out in McClintock's narrative, revising their opinions of Admiralty decisions and showing Sir John Franklin in a wonderfully flattering light. As *Fraser's Magazine* concluded, "[T]o our children's children will the stirring story be told, that Sir John Franklin was the first discoverer of a Northwest Passage."

Although McClintock was physically responsible for discovering the Victory Point Record, the conclusive evidence that led to Franklin's redemption, it was universally acknowledged that Lady Franklin, England's own Penelope, was the force behind the discovery. Lady Franklin's final search expedition (1857–59) to the area north of the Back River was the only expedition – government or otherwise – to discover definitive evidence of the expedition's fate.[90] That the dubious reward for her efforts – the document of her husband's death and the fate of his expedition – was later celebrated by Parliament, the Admiralty, and the press is a testament to the force of her character and her strength as a writer. For "the zeal and self-sacrifice with which she had maintained the search for the missing ships," Lady Franklin was granted the Royal Geographical Society's founder's medal in 1860.[91] Echoing the other dailies and periodicals that carried the story of McClintock's return and his triumphant conclusions, the *Athenaeum* placed honour where honour was most due, on Lady Franklin, who had borne more responsibility than a government had been prepared to bear. It gave thanks specifically to those intangible qualities that made Lady Franklin and her cause so compelling: the "feminine courage which no disaster could dismay, which arose above difficulties as a bird rises above the earth – which neither dulled nor wearied even when strong men grew faint and dubious."[92] In offering her its congratulations, the periodical paid homage to Lady Franklin's faith and perseverance while celebrating her grace, those characteristics of hers that inspired presidents, parliaments, captains, and the public to believe in honour even when it was hidden.

More so than the press editorials and the official documents against which Lady Franklin frequently competed, Lady Franklin's letters, when read together and chronologically, offer a timeline and, paralleling that, an emotional record of the search for Sir John Franklin. From her offer of reward to English whalers in 1848, in which the strength of her hope translated into confidence in her husband's continued survival, to her heartbreaking self-identification as "the widow of Sir John Franklin," to her indebtedness to rising politician Benjamin Disraeli for his support in Parliament, Lady Franklin took her readers on a painful journey through hope, devastation, and faith in a way no other writer on the subject could or did. Her social and political positions certainly aided her in her ability to speak so directly to the Arctic establishment, but it is more her personal strength and obvious intelligence that attract readers today. Lady Franklin had her facts straight, and she used them to her advantage, but her knowledge together with her engaging personality made her letters irresistibly persuasive, to officials and the general public alike.

But while Lady Franklin's letters offer much that is appealing, they also challenge us to read them carefully and critically. Her ethnocentric dismissal of Inuit testimony, her belief in England's cultural and scientific superiority, and her preoccupation with an Englishman's "duty" cannot go unnoticed. These beliefs, as difficult as they may be to accept in a woman so otherwise admirable, nevertheless make Lady Franklin human. Admittedly, I reached this conclusion following a long interrogation of my own assumptions about Lady Franklin and what sort of person I was demanding her to be. Even while I rolled my eyes at the *Leisure Hour*'s efforts to "sing the praises of the Penelope of England," I had to admit that I too was caught up in the desire to make Lady Franklin a symbol. While the Victorian press transformed her into an icon of loyalty, I gave her the mantle of late-twentieth-century morality. Because Lady Franklin challenged the strictly male world of Arctic navigation and single-handedly took on the Admiralty in a public battle (which she clearly won), I assumed she would naturally challenge other cultural boundaries. But such was not the case. I came to realize that my discomfort with many sections of her letters actually helped me in the preparation of this collection. I had hoped to let Lady Franklin speak for herself, and my inability to accept some of what she was saying forced me to leave her to her own devices. Indeed, part of what remains so admirable about Lady Franklin is her refusal to compromise – with the Admiralty, with Parliament, or with her editors and readers.

In reading her letters, one has the opportunity to get a sense of Lady Franklin that secondary sources could never provide. Much of what has been written about her more recently makes her even more difficult to understand. Pierre Berton describes her as "the indomitable Jane" – a nickname that would most likely have horrified her – while Frances Woodward describes her as more demurring than the reader might expect.[93] Lady Franklin was, of course, somewhere in between these two descriptions. Certainly, in reading her letters, one has difficulty discerning how much Lady Franklin really wished to protect her privacy; her lifelong sense of adventure and her forceful personality belie what might be in Woodward's case a romantic rewriting of Lady Franklin's vitality. Writing letters, in fact, was not her first choice of action. Indeed, although she consistently claimed to have "implicit confidence" in her navigators' "enlightened judgment" of Arctic routes and travel techniques, in 1849 she applied to the Admiralty to travel to the Arctic herself so that she might conduct her own small searches from a Hudson's Bay Company fort and be on hand to care for her husband, should he be found.[94]

I hope that in reading the letters in this collection, the reader will come to appreciate all aspects of Lady Franklin's character. Her imperialist, heroic, and vulnerable sides all deserve attention. While the letters often seem to be written by completely different people and are occasionally even mutually contradictory, it is the multi-facetedness of her personality evident in the letters that makes her the fascinating historical figure that she is. As readers, we have to approach her writing on a number of levels. Collectively, Lady Franklin's letters are an important testament to Britain's historical confidence in its ability to succeed at whatever it undertook, to the unrealistic moral responsibilities Britain placed on its national heroes, to one woman's determination to be heard over a cacophony of male voices, and to a wife's devotion to her missing husband – all this and more. We should not be intimidated by their breadth. I sincerely hope that, in reading this collection, as Lady Franklin herself urged in 1856 (document 24), "however it may pall on the ear of some, who apparently judge of it neither by the heart nor by the head, you will not on that, or on any light ground, hastily dismiss it."

ON READING THE DOCUMENTS

This book does not claim to be the first publication to include Lady Franklin's letters. Arctic enthusiasts and readers interested in the history of exploration in what is now Canada will likely already be familiar with Lady Franklin's writing and her involvement in the international search for her missing husband. Pierre Berton, for one, uses her letters extensively in his compendious chronicle *The Arctic Grail: The Quest for the North West Passage and the North Pole, 1818–1909* (1988), and these same letters are frequently quoted in the numerous explications of Sir John Franklin's fate.[1] Writers of Arctic history implicitly pay homage to Lady Franklin and her influence by frequently citing her writing, and through these references she has remained at the centre of a rich and complex narrative. But while Lady Franklin's involvement in the search for her husband is celebrated in nearly every text on nineteenth-century northern exploration, her letters have not yet appeared in their own collection to stand beside the other journals of adventure. This small book is intended to allow Lady Franklin to stop speaking in fragments, and it thereby gives her letters the forum they deserve.[2]

The letters I have decided to include in this collection have one thing in common: Lady Franklin wrote them in private with a view to inciting their recipients to public action. In my research on Lady Franklin and her activities during the period of the search for her husband's expedition (1848–60), I have been particularly interested in Lady Franklin's use of her name and identity to influence powerful politicians, and for this reason I have chosen to include only signed letters written to decision-makers at a national or international level. While Berton writes that Lady Franklin and her niece Sophia Cracroft "bombarded the press with anonymous letters" to apply pressure on the government and keep national interest in her husband high, I have not included any anonymous letters, as it is my belief that they served a different purpose in

Lady Franklin's campaign to save her husband.[3] The unsigned letters were for the general public and were intended, as Berton notes, to maintain the public's interest. They were undoubtedly effective, but it was not through them that Lady Franklin exercised her main influence. Rather, Lady Franklin reached her primary audience with the letters she wrote to the men safely ensconced behind the doors of the Admiralty offices at Whitehall or within the walls of Parliament. While public interest was certainly a factor in the continuation of the searches, the government ultimately held the power to encourage or destroy her hopes. The letters in this collection are thus examples of Lady Franklin's direct appeals to national and international leaders to grant her requests. They trade on their writer's identity and are detailed, pointed, and precise.

One of the points I hope I have made through my selections is that these letters are public documents. If some of them have been "published" before (i.e., in Victorian government pamphlets or in fragments in twentieth-century historical chronicles), it is because they remain significant *as public documents*. It should thus be noted again that this collection, technically, is not the first "publishing" of Lady Franklin's letters, but rather a first collection of them. As readers will notice in the headnotes that accompany each letter (document), many letters are reprinted from *Sessional Papers* of the British House of Commons. In their original form, they were submitted as evidence in adjudications involving members of the Arctic Select Committee, were written in response to a particular policy adopted by the Admiralty, or comprised a series of unsolicited requests that Lady Franklin frequently made to the Lords at Whitehall. Many of these letters are technically "available" to readers because they exist (in whole or in part) in published form, but it is a time-consuming endeavour to find them in their entirety, scattered as they are throughout a mountain of material on the subject. Now that these letters have been uncovered from their various hiding places, it seems only logical that they should form their own collection as a unified opus. As nineteenth-century "public" documents (the *Athenaeum* regularly reviewed Admiralty reports in its book reviews and weekly interest columns), Lady Franklin's letters were, and should continue to be, available to interested readers.

Because Lady Franklin was so prolific in her appeals, often writing similar letters to several people, this collection comprises only a representative selection of her writing. The selected letters offer a timeline of Lady Franklin's opinions and efforts as the search for her husband evolved. They come from various collections, which has made their collation even more of an adventure and enriches the picture they provide of Lady Franklin. While the *House of Commons Sessional Papers* are the source of many of the most recognizable letters, several other archives and collections contain letters of equal and distinct – and more personal – interest. The letters not found in the *Sessional Papers* I have transcribed from collections in England's Public Record Office, Cambridge University's Scott Polar Research Institute, and the microfilm collection of the Hughenden Archives (the archives of Benjamin Disraeli) at Queen's University in Kingston, Canada. Lady Franklin's letters to Disraeli are a unique thread in this collection, as her relationship with the future prime minister was long-standing and personal. The D'Israelis were family friends and belonged to the same book club; Lady Franklin and Disraeli's sister Sarah were close.[4] Regrettably, Lady Franklin's personal connection to Disraeli did little to spur the budding parliamentarian to deliver any remarkable oratory on the subject of her missing husband, but her letters to him reveal a candour that is noticeably absent in her more tailored letters to men she knew only through the quest for her husband. While technically these letters to Disraeli would be considered "private," I include them here because they served the same purpose as those written to the First Lord of the Admiralty or to the president of the United States: they encouraged their recipient to public action.

While it would be interesting to supplement Lady Franklin's public documents – such as the letters in this collection – with her private reflections, this is not always possible, and it is not the point of this book. Frances Woodward admits that there is little in the way of personal reflection from the period of the heaviest searches; we are most likely to know of Lady Franklin's thoughts and fears only through the recorded conversations and correspondence of other people and through her letters to others. Woodward claims that only in 1855 did Lady Franklin begin to keep a regular diary, but she considers it "such a diminutive affair that it is very difficult to decipher it; nor is the task a rewarding one."[5] Ken McGoogan elaborates: "Even in her diaries and notebooks, Jane Franklin recorded little that could be called personal. When she strayed from her 'scientific' approach and put too-revealing sentiments on record, often she later went back … and excised passages."[6] Lady Franklin's desire to record the details of her life and world was outweighed by her need for privacy, and she is known to have destroyed "entire notebooks" from precisely those periods during which her private insights would have proved most interesting.[7] While historians are frustrated by the dearth of personal comment in Lady Franklin's writing, her public comments even at the height of the searches were few and far between. Lady Franklin was in fact often unavailable for comment when her cause was being discussed within the walls of Parliament or in the Admiralty offices at Whitehall. She filled her time with activities that kept her mind off the subject that otherwise preoccupied her while in London. According to John Knox Laughton, Lady Franklin "sought relief from oppressing anxiety in constant travel" during the years of the searches, and in late 1859 she was abroad when her own schooner the *Fox* (which carried the expedition responsible for having "cleared up the mystery") docked back in England.[8]

I did not select the letters included here according to their level of influence and success. Indeed, many of them produced little response or received a firm answer in the negative. I was more interested in the letters themselves for the story they convey of her efforts. Lady Franklin's correspondence was varied, articulate, and virtually unceasing, and I would argue that the letters in fact got better as they became less successful. Unsurprisingly, her efforts to articulate the exigency of the situation and to convince her readers of the need to continue the search were inversely proportionate to the official support she received, although, as she herself admitted, she was unable to resist pressing for more support even when the Admiralty did supply her with ships, men, and stores (document 16). Her rhetorical vacillations between encouragement and admonishment, and her journey from hope to anger and from anger to reconciliation, can be seen only when the "successful" letters are read together with the "failures."

Thus, I have found it most useful to present the letters chronologically, rather than according to their immediate influence, as this allows the reader to notice the abrupt change in tone across a distinct temporal divide. Lady Franklin's letters in this collection can be divided into two periods: 1848–53 and 1854–60 – before the announcement of her husband's death and after. The Admiralty's January 1854 decision to consider

Franklin and his men as “having died in Her Majesty’s service” and Dr John Rae’s 29 July 1854 report of English sailors having starved to death near the Back River mark the transition between the two periods. In the first defined period, Lady Franklin wrote as the wife of her “absent husband.” Her appeals to the Admiralty during this time reveal a faith both in her husband’s ability to survive and in the Admiralty’s ability to save him, if given proper encouragement. Lady Franklin’s later letters noticeably change tack. Weighed down initially by the “crushing misery of Dr Rae’s report,” after January 1854 Lady Franklin continued to plead her husband’s cause, hoping that he might be found, if not alive, then at least dead in heroic endeavour. She identified herself during this period as “the widow of Sir John Franklin,” and instead of encouraging the Admiralty Lords to do what they did best – send valiant men out to battle with the elements – she began to berate them for their perfidy in their treatment of her husband (documents 3, 19).[9] In her letters throughout both periods, however, whether in moments of hope or pessimism, Lady Franklin continued to express the belief that something would be discovered, that her husband would be redeemed. It is Lady Franklin’s enduring faith that makes these letters so powerful.

FORMAT OF LETTERS AND ARTICLES

The format of the original letters has been retained to the greatest degree possible, given the obvious difficulties in transforming the handwritten page into typescript. The majority of Lady Franklin’s letters are impressively long, and it was not feasible to reproduce the individual pages of each. Original page breaks are acknowledged by an oblique (“/”) in the text. All other internal idiosyncrasies, such as indentations, abbreviations, additions, italics, underlining, deletions, and the use of an apostrophe (“ ‘ ”) in printed documents to indicate a “ ^{c} ” (as in McClintock), have been faithfully retained. Although this may occasionally make a letter difficult to read, I felt that a presentation that closely resembled the author’s original writing would give the reader the experience of reading the originals and perhaps convey a truer sense of the letters and of the personality and thought processes of their author than a modernized format might.

Headnotes of Letters

The headnote of each letter contains all of the identification information:

Addressee: Name is given in square brackets ([ ]) if not explicit in letter.
Place from which letter was sent
Date of letter
Source of manuscript
Heading: Title given to the document by the source. Punctuation and capitalization within the heading are transcribed from the source.
Editorial Comment: Any information that is pertinent to the context of the letter or its contents, such as spelling errors, dating techniques, etc.

Endnotes of Letters

I have added endnotes wherever they might prove helpful, but given the detail and eloquence of Lady Franklin's writing, I have also attempted to let the letters speak for themselves. Many of Lady Franklin's letters contain internal contextualization, and so issues or concepts may be cross-referenced within a letter for clarity. Any discrepancy in information has been endnoted. Information in each endnote is referenced (see Bibliography), and if the reader requires additional information on a particular subject, it can easily be found in the source indicated. I have retained Roman numeral citations for *BDL* (*Benjamin Disraeli Letters*), H (Hughenden Archives), *HCSP* (*House of Commons Sessional Papers*), and *HLSP* (*House of Lords Sessional Papers*) in order to make use of pre-existing internal reference systems.

Spelling Errors

The original letters and articles in this collection have very few errors. Considering the haste in which some of them were written, the length of some, as well as the fact that mistakes could not be erased, this is truly impressive. Those words that have been misspelled have been left in their original spelling and are noted in the headnote under Editorial Comment by "[*sic*]." Words and terms that were spelled differently in Lady Franklin's time than today (such as Esquimaux, Behring['s] Strait, and inclose) retain the original spelling and are not acknowledged under the "[*sic*]" category.

Author Insertions and Deletions

Where Lady Franklin herself has edited a document and inserted a word or symbol above her original text, I have included it in the line in angle brackets: "<>." This is to minimize disruption in the text with too many superscript words and to remain true to Lady Franklin's intentions. Where she has crossed out a word or symbol, if legible I have included it in the transcription with a strikethrough line: ~~deletion~~.

Abbreviations

The punctuation of abbreviations has been standardized, though superior lettering has been retained. Frequently in the collection, a handwritten superior letter or group of letters has periods beneath it. When this is the case, the period has been replaced by an underline, as in "M$^{\underline{r}}$" or "2$^{\underline{ndly}}$." Only those superscript letters that were marked with inferior periods have been underlined.

Punctuation

Except for the standardization of superior punctuation (see above), no punctuation has been changed in this collection and very little has been added. By contemporary standards, there is perhaps a surfeit of commas, but this is recognized as a particular style. Sometimes the complexity of Lady Franklin's sentences demands what we might today consider heavy punctuation. The original punctuation is never omitted and punctuation has been added only to maintain consistency in the text; additions are indicated by the use of square brackets "[]." Apostrophes are also infrequent in these letters, and I decided not to add any where they were missing, as the additions, indicated by "[]," would break up the continuity of the letters. Words missing apostrophes are acknowledged in the Editorial Comment with "[*sic*]."

CHRONOLOGY OF EVENTS, 1845–1859

The chronology presented on the following pages includes the events that transpired around the search for the lost Franklin expedition that are mentioned in the documents in this collection. The corresponding documents of Lady Franklin, identified by number, appear in the margins so that readers can see where they fit in the larger context of the searches.

One of the striking features of the Franklin searches was their multi-faceted nature. All of the searches had as their primary mandate the rescue and succour of the missing explorers or the discovery of their fate, but all were engaged as well in active exploration of the remaining blank spaces on the Arctic map – perhaps to the detriment of the search. These dual responsibilities are acknowledged here. The search for Sir John Franklin was the necessary catalyst – or excuse – for further exploration of the Arctic, and within only a dozen years, the search expeditions nearly completed the map of northern Canada.

Alan Cooke and Clive Holland's *Exploration of Northern Canada* (1978) offers an astoundingly complete record of exploration in the North from 500 to 1920, and thus provides a rich historical context for the flurry of activity between 1845 and 1860 evident in the documents presented here. Unless otherwise noted, the information in the Chronology comes from Cooke and Holland's *Exploration*.

	1845	On 19 May, commanding the *Erebus* and *Terror*, Captain Sir John Franklin and 128 officers and men enter the Arctic to pursue the British Navy's most ambitious attempt to discover the Northwest Passage. Their ships are last seen anchored to an iceberg, waiting for safe entrance into the eastern Arctic.
	1847	On 19 February, letters of enquiry are sent to British Arctic explorers and experts asking their opinions on Franklin's possible whereabouts. The Admiralty declares its "unlimited confidence in the skill and resources of Sir John Franklin."[1]
	1847	In the whaler *St Andrew*, Captain William Penny[2] attempts to enter Lancaster Sound "with the dual object of whaling and searching for Sir John Franklin's missing expedition."[3] Bad weather and rough ocean force him to abandon the search.
	1847–49	Seasoned British explorer Sir John Richardson[4] and Hudson's Bay Company surveyor Dr John Rae[5] conduct an overland search for Franklin north of the HBC outposts; they find no trace.
document 1	1848	20 March, Lady Franklin offers her first reward for the rescue of her missing husband. Her £2,000, together with 100 guineas from the Admiralty, is intended to gain information. The reward is advertised to British whalers.

	1848–49	Captain Sir James Clark Ross[6] reluctantly commands a two-ship expedition to the eastern Arctic that is meant to act in tandem with the overland search of Richardson and Rae. The two expeditions never meet; Ross does not leave England until mid-June, and by the time he arrives in the Arctic, freeze-up has begun and his ships are unable to penetrate the ice.[7] Ross is forced by adverse circumstances to return to England early.
	1848–50	Captain Henry Kellett[8] of the *Herald* makes three summer surveys of the western Arctic in search of Franklin; he finds nothing.
	1848–52	In the *Plover*, Captain Edward Laws Moore[9] searches for Franklin in the western Arctic through Bering Strait, overwintering in 1848 on the Asiatic side. The expedition remains in the North for three years, the crew conducting searches in smaller boats.
	1849	On 17 January, the first official meeting of the Arctic Council convenes, comprising "Officers, &c. interested in the welfare of the Arctic Expeditions."[10]
documents 2, 3	1849	On 23 March, Lady Franklin prints her second private announcement of reward for British whalers, this time offering £3,000 to any ship that travels beyond its whaling grounds to search for her husband. This reward, printed in broadsheet and distributed by Lady Franklin to whalers in the Orkneys,

accompanies advance notice of a £20,000 reward to be offered by the Admiralty.

1849 The Admiralty's £20,000 reward for discovery of the missing ships is announced in Parliament, but it is considered too late in the season to equip a navy expedition for an extended search of Franklin's supposed route.

1849 In the early summer, the Admiralty supply ship *North Star* sails to the Arctic with provisions for Ross and Franklin. It arrives in the Arctic, drifts "hopelessly about in the pack for sixty-two days," and is forced to winter in the ice, unable to deliver its supplies.[11]

1849 Captain Penny, now of the whaler *Advice*, commands a two-ship summer whaling and search expedition near Baffin Bay and into Lancaster Sound. As they sail out of the Arctic, the crews leave caches of supplies. Upon his return to England, Penny offers his services to the Admiralty to participate in the official search.

documents 4, 5

1849 Inspired by Lady Franklin's appeal to his government on behalf of her husband, US president Zachary Taylor attempts to organize an American rescue expedition to the Arctic.[12] In May, the *Times* announces the American Government's intention "to send two ships in search of Sir James [*sic*]

		Franklin's expedition, one to go round east by Labrador and Davis's Straits; the other round west by Behring's Straits."[13]
documents 6, 7, 8	1849	After Richardson's return to England, Dr Rae continues an overland search for Franklin but is unable to cross from the mainland to Wollaston Peninsula because of ice conditions. He turns back only about eighty miles from where Franklin's ships are beset in the ice.
documents 9, 10	1849	After failing to secure funding and approval for a government-supported search in the 1849 season, American president Taylor distributes notices of the British reward and recommendations for searches to American whalers and sailors.[14]
documents 11, 12	1850	On 4 January, President Taylor publishes for Congress his correspondence with Lady Franklin in an attempt to garner support for an American expedition.
	1850	On 8 March, the *Gazette* announces the Admiralty's reward of £20,000 to any expedition able to rescue any or ascertain the fate of the missing explorers.
	1850–51	Captain Horatio Thomas Austin[15] heads a four-ship expedition into the eastern Arctic, victualled for three years. Austin's expedition comprises two

steam-powered vessels, the *Pioneer* and the *Intrepid*, and two sail-powered vessels, the *Resolute* and the *Assistance*. In the words of Sherard Osborn, captain of the *Pioneer*, the sail-powered ships resemble "ungainly snuffboxes; … their bows [forming] a buttress which rather [pushes] the water before it than [passes] through it," and offer little hope for an expeditious search.[16] Captain Erasmus Ommanney,[17] Austin's second-in-command and captain of the *Assistance*, has "the satisfaction of meeting with the first traces of Sir John Franklin's expedition" – evidence of a brief encampment on the south of Devon Island.[18] The expedition returns home after only one year.

1850–51 This time sailing under the Admiralty flag, Captain Penny commands a two-ship expedition to the same area being searched by Austin's expedition (Wellington Channel). Conducting sledge journeys from the frozen-in *Lady Franklin* and *Sophia*, Penny's crews discover Franklin's first winter quarters and the three marked graves on the shore of Beechey Island, just south of Devon Island. Unable to procure a steamship from Austin to continue exploring Wellington Channel, Penny is forced to return to England after one winter.[19]

1850 Captain Charles Codrington Forsyth[20] commands the *Prince Albert*, Lady Franklin's first privately funded expedition, on its first voyage into the Arctic. The ship remains out four months, unable to enter the ice, and returns to

	England with a note from Ommanney describing the find of Franklin's 1845–46 winter quarters on Beechey Island.
1850	On 2 May, a joint resolution in the US Congress allows "the President of the United States to accept and attach to the navy two vessels offered by Henry Grinnell, esq., of New York,[21] to be sent to the Arctic seas in search of Sir John Franklin and his companions." The ships, to be outfitted at Grinnell's expense, are to be equipped for three years in the Arctic.[22]
1850–51	Financed jointly by the Hudson's Bay Company and public subscription, veteran Arctic explorer Sir John Ross[23] commands a two-ship expedition to the eastern Arctic in the *Felix* and *Mary*. The expedition winters near Penny's and Austin's expeditions and helps to explore Beechey and Cornwallis Islands.[24]
1850–51	Under the direction of the Hudson's Bay Company, Dr Rae conducts a coastal search around the mouth of the Coppermine River and Victoria Island. On its southern shore (in Parker Bay), Rae discovers two pieces of wood which he identifies as having come from a British ship.[25]
1850–51	Henry Grinnell's ships join the search for Franklin. Commanding the *Advance* and the *Rescue*, Captain Edwin Jesse De Haven[26] attempts to enter Wellington Channel to search for Franklin, but he is unsuccessful. The

ships are "caught and frozen in the ice and for nine months [are] confined in this perilous situation, drifting to and fro more than a thousand miles."[27] Eventually the ships are pushed out of the mouth of the channel into Lancaster Sound and finally out into Baffin Bay.[28]

1850–55 Captain Richard Collinson[29] commands a two-ship expedition to the western Arctic in the *Investigator* and *Enterprise*. Before reaching Bering Strait, Collinson loses contact with his second-in-command, Captain Robert John Le Mesurier McClure,[30] and spends the next five years searching for both him and Franklin. Having never received official sailing orders from his superior officer, McClure enters the western Arctic with the dual ambition of locating Franklin and a passage. In a disturbing echo of Franklin, McClure communicated to the Admiralty on his departure that "no apprehension whatever need be entertained of our safety until the autumn of 1854, as we have on board three years of all species of provisions."[31] After being stuck in the ice for three winters, suffering acutely from scurvy and on the verge of starvation, McClure's remaining crew members are rescued in late 1853 by a sledge party from the *Resolute* (commanded by Captain Kellett), part of an 1852 search expedition from the eastern Arctic. On their trek to safety, the men pass over a frozen Northwest Passage. Meanwhile, following traces of the *Investigator*'s path but needing to husband his own resources, Collinson stays in the Arctic for five years, unable to locate either McClure or Franklin. He sails east along the northern coast of North America and reaches within eighty miles of Franklin's ships in 1851 and again in 1852. While, like McClure, he is unable to sail through the central Arctic to the east, Collinson too discovers a Northwest Passage and, in spite of treacherous conditions, has great sailing success.

1851–52 The *Prince Albert* once more journeys north, again sponsored by Lady Franklin and now under Captain William Kennedy.[32] Lady Franklin wrote the sailing orders herself and provisioned the expedition for two years. Unable to penetrate the ice and hearing news of another, larger, government-sponsored expedition, Kennedy returns home after only sixteen months, having found no trace of Franklin.

1852 Lady Franklin's second ship, the *Isabel*, is sent to Jones Sound, high in the eastern Arctic, and then along the coast of Baffin Island, under the command of Captain Edward Augustus Inglefield.[33] The ship reaches Baffin Island but returns home at the onset of winter with no new discoveries.

1852–54 Commodore Sir Edward Belcher[34] pilots a five-ship expedition, including the *Intrepid* under Captain Leopold McClintock,[35] intended to last four years – the "last" search for Franklin. The expedition is plagued by disagreements between the captains and a prevalent feeling of foreboding. Late in 1853, Belcher's men rescue McClure's crew, but they find no sign of Franklin.

As soon as is possible in 1854 – and after having received orders from the Admiralty to return to England – the expedition abandons four of its five ships and returns home.

document 13

1853 Captain William Kennedy sails Lady Franklin's yacht the *Isabel* to Valparaiso, Chile, with the intention of entering the western Arctic. The expedition is abandoned at Valparaiso when the men mutiny and the Admiralty refuses to supply new personnel.[36]

1853–55 American entrepreneur Henry Grinnell launches his second privately funded expedition, under the command of Elisha Kent Kane,[37] in the *Advance*. The ship is beset by ice on the west coast of Greenland for two years. In the summer of 1855, the crew members abandon the ship, eventually meeting with an American relief expedition sent to their aid under the command of Lieutenant Henry J. Hartstene.[38]

1853–54 John Rae travels overland to complete a "survey of the northern coast-line of the North American mainland."[39] In the spring of 1854, he receives a report from Inuit hunters that a party of European sailors died of starvation on King William Island, eventually succumbing to cannibalism "as a means of prolonging existence."[40] Rae hastens to England to deliver his report to the

Admiralty and to stop any further expenditure and "save the risk of more valuable lives being jeopardised in a fruitless search" for Sir John Franklin.[41]

1853 In October, Captain Inglefield returns to England from his summer supply run with the news that Captain McClure has discovered the Northwest Passage. On board he carries Samuel Cresswell of McClure's ship, the HMS *Investigator*; Cresswell is the first person to travel through the Northwest Passage back to England. In December, the *Times* states confidently that the search for Franklin is over.

documents 14, 15, 16, 17, 18

1854 On 19 January, a notice appears in the *Gazette* stating that "[n]otice is hereby given, that if intelligence be not received, before the 31st March next, of the officers and crews of Her Majesty's Ships 'Erebus' and 'Terror' being alive, the names of the officers will be removed from the Navy List, and they and the crews of those ships will be considered as having died in Her Majesty's service."

1854 In October and with the support of the Arctic Select Committee, Captain McClure receives £5,000 for the discovery of the Northwest Passage. According to the committee formed to adjudicate the award, the "evidence [given to the committee places] beyond doubt that to Captain M'Clure incontestably belongs the distinguished honour of having been the first to perform the actual passage over water along the Northern Coast of America."[42] The same amount is divided among his surviving crew members.

documents 19, 20, 21 | 1855 | Hudson's Bay Company chief factors James Anderson[43] and James Stewart[44] conduct a canoe trek to the site of Rae's report; the canoes are not fit for, and time does not permit, an ocean crossing of Simpson Strait to King William Island. Unknown to Anderson and Stewart, three men from the expedition sight a ship in the distance but do not report their discovery.[45] Because of a shortage of supplies and the state of the canoes, Anderson and Stewart are forced to turn back without having reached the area delineated by Rae's report. Instead, they explore the nearby Montreal Island, collecting relics and communicating through sign language with Inuit people. Their findings satisfy the Admiralty as to the veracity of Rae's report.[46]

1855 | In September, one of Belcher's four abandoned ships, the *Resolute*, is discovered attached to an iceberg by the American whaler *George Henry*. Having travelled 1,200 miles and still seaworthy, the *Resolute* is boarded and sailed back to the United States, where it and all its contents are given as a gift from the English government to the crews of the *George Henry*. In an unanimous decision, the American Congress votes to provide $40,000 for the *Resolute*'s purchase and refit for continued Arctic service, and to return it to England as a gift to Queen Victoria.

document 22	1856	On 22 January, the Admiralty announces in the *Gazette* that "Dr Rae having claimed to be entitled to the reward of £10,000, under the terms of the [1850 Reward] Proclamation, they will proceed within three months from the date hereof to adjudicate on such claim; and that all such persons who, by virtue of such Proclamation, deem themselves entitled to the whole or any part of the reward in question, must prefer their claims within such time, after the expiration of which no claim will be entertained."[47] On 16 February, however, the Admiralty orders a draft of the final reward notice, "stating that adjudication of the £10,000 will be made as soon as my Lords receive the report of [Anderson and Stewart's] expedition," suggesting that the decision to acknowledge Rae as the discoverer of Franklin's fate is already determined.[48]
document 23	1856	On 24 June, the Admiralty announces its decision to award Dr Rae £10,000 for ascertaining the fate of the Franklin expedition.[49] Despite public interest in equipping a final expedition to the Back River, the Admiralty makes no further plans to complete the search and verify Rae's findings.
document 24	1856	A memorial is delivered to Viscount Palmerston, British prime minister,[50] by "men of science and others who have taken a deep interest in Arctic discovery, or explorers who have been employed in the search for our lost countrymen," urging the government to outfit one final search in the area delineated by Rae.

documents 25, 26	1857	The *Resolute* arrives back in England, a gift from the American government, to resume the search for Franklin. The British government determines against mounting a final search and the *Resolute* is abandoned.[51]
	1857–58	Captain Penny winters in Cumberland Sound, Baffin Island, combining long-term whaling with a continued search for information on Franklin, even though he is far from the identified area of the search.[52]
documents 27, 28	1857–59	In April 1857, refused the use of an Admiralty ship, Lady Franklin funds her last private search expedition, in the *Fox*, commanded by Captain Leopold McClintock.[53] Though initially reluctant to participate, the navy supplies much of the provisions for the expedition, including all powder and shot for hunting, powder for ice-blasting, ice gear, hydrographical instruments, winter housing, and 6,682 pounds of pemmican.[54] The expedition leaves England in early July, and even before entering Baffin Bay, it encounters a well-formed ice pack; by the middle of August, the *Fox* is able to advance only "a ship's length" each day, on average.[55] By 12 March 1858, the *Fox* is pushed "ingloriously out of the Arctic regions" by the south-flowing pack, but eventually frees itself and returns for another season.[56] This time, travelling directly to the shores of King William Island, McClintock discovers the skeletons of members of Franklin's crew. On a sledge journey to the west coast of King William Island, Lieutenant William Robert Hobson[57] discovers a single sheet of paper under a

cairn. Recording the date of Franklin's death in 1847, the paper clears his name from the ignominy of cannibalism and records the intended abandonment of the *Erebus* and *Terror* by remaining crew members in 1847.

document 29 1860 On 2 March, notice is put forward in a cheering Parliament for a reward for Captain McClintock for his "singular expedition" and the "noble lady through whose energy and inspiration that expedition was so gallantly undertaken and successfully accomplished."[58] McClintock is awarded £5,000, and money is voted for a monument of Franklin to be erected at Waterloo Place, acknowledging him and his crew as the first discoverers of a Northwest Passage. To paraphrase Franklin's early travelling companion Sir John Richardson, "They forged the last link with their lives."[59]

Daguerreotype of Sir John Franklin aboard *Erebus*, 1845.

21 Bedford Place, Lady Franklin's old family home, now the Penn Club, a semi-private hotel established shortly after World War I. Photo courtesy of John and Eileen Elce.

(*above*) 33 Spring Gardens, now the back of Trafalgar Studios (formerly the Whitehall Theatre). Photo courtesy of John and Eileen Elce.
(*below*) Admiralty offices, facing Whitehall. Photo courtesy of John and Eileen Elce.

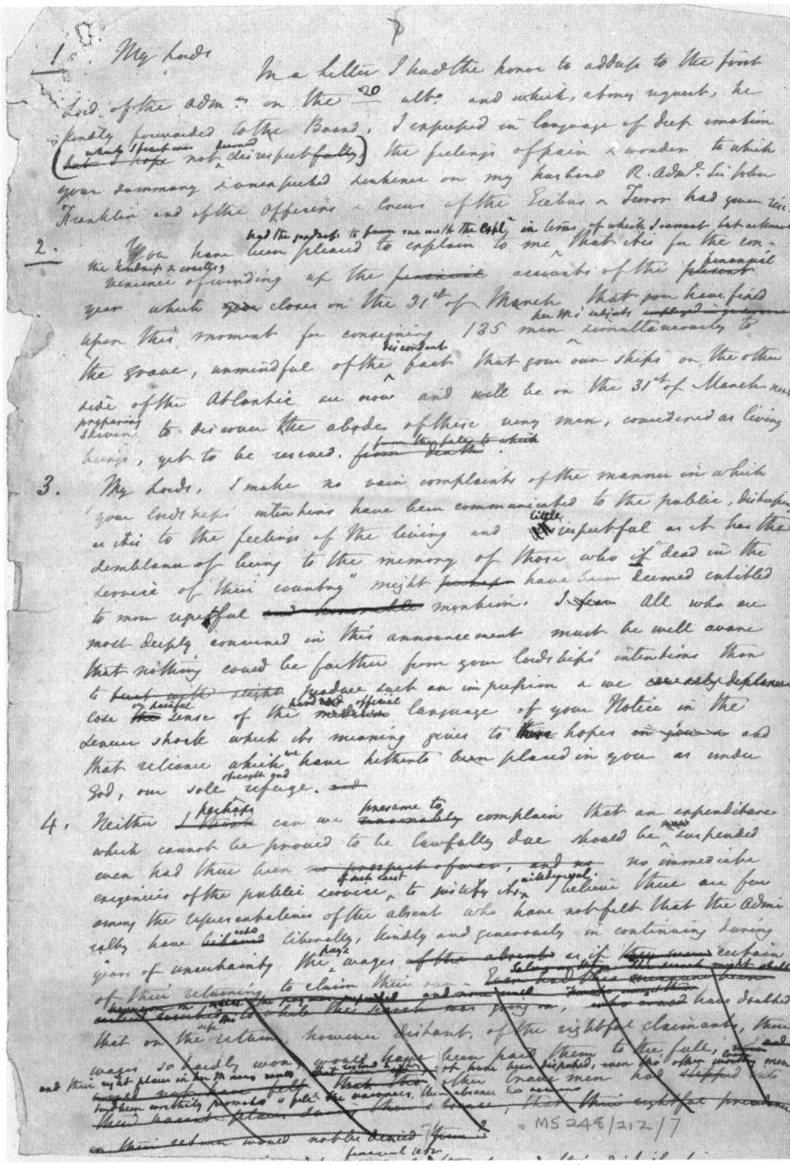
1. My Lords

In a letter I had the honor to address to the first Lord of the Adm. on the 20 ult. and which, at my request, he kindly forwarded to the Board, I expressed in language of deep emotion (which I trust was deemed not disrespectful) the feelings of pain & wonder to which your summary & unexpected sentence on my husband R. Adml. Sir John Franklin and of the Officers & Crews of the Erebus & Terror had given rise.

2. You have been pleased to explain to me that it is for the convenience of winding up the financial accounts of the present year which closes on the 31st of March that you have fixed upon this moment for consigning 135 men simultaneously to the grave, unmindful of the fact that your own ships on the other side of the Atlantic are now and will be on the 31st of March next [illegible] to discover the abode of these very men, considered as living beings, yet to be rescued.

3. My Lords, I make no vain complaint of the manner in which your lordships' intentions have been communicated to the public, distressing as it is to the feelings of the living and little respectful as it has the semblance of being to the memory of those who if dead in the service of their country might have been deemed entitled to more respectful mention. All who are most deeply concerned in this announcement must be well aware that nothing could be farther from your lordships' intentions than to produce such an impression & we deeply deplore [illegible] sense of the [illegible] official language of your Notice in the severe shock which its meaning gives to hopes [illegible] and that reliance which we have hitherto been placed in you as under God, our sole strength and refuge.

4. Neither can we presume to complain that an expenditure which cannot be proved to be lawfully due should be suspended even had there been no immediate exigencies of the public service, to justify it. I believe there are few among the well-wishers of the absent who have not felt that the Admiralty have acted liberally, kindly and generously in continuing during years of uncertainty the wages [illegible] of their relatives to claim their [illegible] that on the return, however distant, of the rightful claimants, their wages so kindly won [illegible] been paid them to the full, [illegible]

MS 248/212/7

Draft of Lady Franklin's 24 February 1854 letter to the Lords of the Admiralty. Scott Polar Research Institute, MS 248/212/7. Used with permission.

Statue of Sir John Franklin in Waterloo Place. Photo courtesy of John and Eileen Elce.

THE DOCUMENTS:
SELECTED LETTERS OF JANE, LADY FRANKLIN

1

REWARD OFFERED BY LADY FRANKLIN[1]

21 Bedford Place, London,[2] [Monday] 20 March 1848

Source: Printed transcript, *HCSP*, 1847–48/XLI/211

Heading: PRINTED DOCUMENT. – REWARD offered by Lady *Franklin*.

Editorial comment: This private announcement was printed by the Admiralty for large-scale distribution to English whalers. At least 200 copies were made: 50 to be distributed to English whalers through customs officials; 50 to go to employees of the Hudson's Bay Company; and 100 to be carried on the 1848 government search expedition to be distributed to any whalers met on the way (*HCSP*, 1847–48/XLI/212).

Two Thousand Pounds offered by Lady *Franklin* to Whalers.

WITH the view of inducing any of the whaling ships with resort to Davis Straits and Baffin Bay, to make efforts in search of the expedition under the command of Sir John Franklin, in those parts which are not within the scope of the expeditions about to be sent out by Government, I hereby offer one thousand pounds (1,000 *l*.), to be divided as follows: to the owner, captain, officers and crew of any ship which shall depart so far from the usual fishing grounds as to explore Prince Regent Inlet, Admiralty Inlet, Jones Sound, or Smith Sound, provided such ship, finding the above expedition in distress, shall communicate with, and afford it effectual relief.

To the owner, two-tenths, or _ _ _ _ _	£. 200
" captain, one-tenth, or _ _ _ _ _ _	100
" chief mate, one-twentieth, or _ _ _ _	50
" next two officers, one fortieth, or 25*l*. each _ _	50
The remaining six-tenths, or to be _ _ _ _ _ divided amongst the rest of the ship's company.	600

And further, I hereby offer an additional sum of one thousand pounds (1,000*l.*), to be distributed in the same proportions to the owner, officers and crew of any ship which shall at an early period of the season make extraordinary exertions for the above object, and, if required, bring Sir John Franklin and his party to England.

The whole or part of this last 1,000*l.* will be granted according to the decision of Sir John Franklin, or the commanding officer of the expedition relieved. In other respects the decision of the following gentlemen, who have kindly consented to act as referees in awarding the above 2,000*l.*, is to be final, viz. Admiral Beaufort,[3] Captain Sir W. Edward Parry, R. N.,[4] Thomas Ward, Esq., Hull.[5]

20 March 1848.[6]

Jane Franklin,
21 Bedford-place, London.

2

To: [BENJAMIN DISRAELI][1]
4 Spring Gardens, London,[2] [Thursday] 22 March 1849
Source: Original, H, A/IV/N/2
Editorial comment: Mary Anne Disraeli has endorsed the back of the ms: "1849 | March 24th | Lady Franklin." "22 March 1849" is also written in the top right-hand corner of the letter in an unknown hand (possibly an archivist's).

Dear Sir

I beg to be allowed to introduce to you my friend the Rev[d] J. P. Gell, late Warden of Christ's College, VDiemen's Land.[3]

If you can spare him a few moments' conversation, he will as briefly as possible explain the nature of the great favor I am desirous of obtaining at your hands. You will excuse my adding that to me this interview is of urgent importance as affecting the fate of my absent husband at a very critical moment.[4]

I am, dear Sir
yours truly
Jane Franklin

4. Spring Gardens
22[d] March 1849

3

LADY FRANKLIN TO WHALERS

[21 Bedford Place, London, Friday 23 March] 1849

Source: Printed broadsheet, NLC, G662 F73 1849 xxfol

Editorial comment: Like Lady Franklin's first offer of reward (see document 1), this broadsheet was meant for general distribution. Lady Franklin, who at this time made a journey to the Orkneys in search of news, may have delivered these broadsheets herself to whaling captains. The reward was meant to act in conjunction with the Admiralty's reward of £20,000, drafted by the Admiralty in order to be available for the 1849 season (*HCSP*, 1850/XXXV/328).

THREE THOUSAND POUNDS,

OR A PROPORTION THEREOF, ACCORDING TO THE SERVICES RENDERED,

OFFERED BY LADY FRANKLIN,

TO SUCH OF THE WHALING SHIPS AS SHALL BE GENEROUSLY INCLINED TO ASSIST THE SEARCH FOR SIR JOHN FRANKLIN AND HIS GALLANT COMPANIONS.

With the view of inducing any Whaling Ships which shall resort to Davis Strait and Baffin Bay to make special efforts in search of the expedition under the command of Sir John Franklin, – I hereby offer the sum of Three Thousand Pounds, (£3000) or a proportion thereof, according to the services rendered, to such ship or ships as, departing from the usual fishing grounds, shall discover, and, if needed, afford effectual relief to the above expedition, or to any portion of it.

It is proposed that the amount of reward, according to the efforts made and services performed, shall be determined by the following gentlemen, who have kindly consented to act as referees, viz:

REAR-ADMIRAL SIR FRANCIS BEAUFORT, K. C. B.,[1]
CAPTAIN SIR W. EDWARD PARRY, R. N.,[2]
THOMAS WARD, ESQ., (HULL.)[3]

In regard to the distribution of the sum awarded, among owners, captains, officers, and seamen, the amount to each to be adjusted in the same proportions as if similar value of produce from the fishery had been obtained.

In the event of more than one ship making special efforts to give succor to the expedition, each ship is to receive its proportion of the reward agreeably to the decision of the referees.

The attention of whalers disposed to aid in this service is particularly directed to the Gulf of Boothia, within Regent Inlet, or to any of the Inlets or Channels leading out of Barrow Strait, or the sea beyond, either northward or southward, as also to any Sounds or Inlets in the north and western sides of Baffin Bay, above the 75th degree of latitude.

Should it be clearly proved and ascertained that any whaler has made extraordinary efforts, or special researches in quarters remote from the ordinary fishing grounds, for the purpose designated, though no success may have attended their endeavors, the case of such whaler, with a view to reward, will be taken into favorable consideration by the referees.

For the greater satisfaction of parties claiming reward, the owners and captains shall be authorized, if they desire it, to nominate one additional referee, who shall act and vote in all respects as the standing referees in the special case for which they may be nominated. The referees being then increased to four, will, according to the usual order of business, choose for themselves a fifth, as umpire.[4]

4

To: [ZACHARY TAYLOR]
[21] Bedford Place, London, [Wednesday] 4 April 1849
Source: Handwritten transcript, H, A/IV/N/5a
Editorial comment: A transcription in an unknown hand. It may be Sophia Cracroft's hand, as she frequently acted as Lady Franklin's secretary. [*sic*]: enterprize. This is the most famous of Lady Franklin's epistles. Published in full in several newspapers (in England and internationally), it was frequently mentioned in Parliament in order to arouse interest in the Arctic search and was described as "one of the most eloquent letters addressed by man or woman to man or woman."

Copy

Bedford Place. London.
4th April 1849.

Sir,

I address myself to you as the head of a great nation, whose power to help me I cannot doubt, and in whose disposition to do so, I have a confidence which I trust you will not deem presumptuous[.]

The name of my husband, Sir John Franklin, is probably not unknown to you. It is intimately connected with the northern part of that continent of which the American Republic forms so vast and conspicuous a portion. When I visited the United States, 3 years ago, amongst the many proofs I received of respect and courtesy, there was none which touched and even surprised me more than the appreciation every where expressed to me of his former services in geographical discovery, and the interest felt in the enterprize in which he was then known to be engaged.[1]

The Expedition fitted out by our Government for the discovery of the North West Passage (that question which for 300 years has engaged the interest and baffled the energies of the man of science and the navigator) sailed under my husband's command in May 1845. The two Ships, "Erebus" and "Terror" contained 138 men (officers and crews) and were victualled for 3 years.[2] They were not expected home unless success had early rewarded their efforts or some casualty hastened their return, before the close of the year 1847, nor were any tidings expected from them in the interval. But when the autumn of 1847 arrived, without any intelligence of the Ships, the attention of H. M. Government was directed to the necessity of searching for, and conveying relief to them, in case of their being imprisoned in ice, or wrecked and in want of provisions and means of transport.

For this purpose, an expedition in 3 divisions was fitted out in the early part of last year, directed to 3 different quarters simultaneously viz 1. to that by which, in case of success, the ships would come out of the Polar Sea to the westward (or Behrings Straits) 2ndly to that / by which they entered on their course of discovery on the Eastern side, (or Davis Strait) and 3rdly to an intervening quarter comprising a portion of the Arctic Shore approachable by land from the Hudson Bay Company's Settlements, on which it was supposed the Crews, if obliged to abandon their ships, might be found. This last division of the Expedition was placed under the command of my husband's faithful friend, the companion of his former travels, Dr Sir John Richardson, who landed at New York in April of last year, and hastened to join his men and boats which were already in advance towards the Arctic Shore. Of this portion of the Expedition, I may briefly say that, the absence of any intelligence from Sir John Richardson at this season, proves that he has been unsuccessful in the object of his search.[3]

The Expedition intended for Behrings Straits has hitherto been a complete failure. It consisted of a single ship, the Plover, which owing to her setting off too late and to her bad sailing properties, did not even approach her destination last year.[4]

The remaining and most important <part> of the searching Expedition, consists of 2 ships under the command of Sir James Ross which sailed last May for Davis Straits, but did not succeed, owing to the state of the ice, in getting into Lancaster Sound, till the season for operations had nearly closed. These ships are now wintering in the ice, and a storeship is about to be dispatched from hence with provisions and fuel to enable them to stay out another year, but one of these vessels is in a great degree withdrawn from active search by the necessity of watching at the entrance of Lancaster Sound for the arrival of intelligence & instructions from England by the whalers.[5]

I have entered into these details with the view of proving that, though the British Government has not forgotten the duty it owes to the brave men whom it has sent on a perilous service, and has spent a very large sum in providing the means for their rescue, yet that, owing to various causes, the means actually in operation for this / purpose are quite inadequate to meet the extreme exigence of the case. For it must be remembered that the missing ships were victualled for 3 years only, and that nearly 4 years have now elapsed, so that the survivors of so many winters in the ice must be at the last extremity, and also, it must be borne in mind that the channels by which the ships may have attempted to force a passage to the westward, or which they may have been compelled by adverse circumstances to take, are very numerous & complicated, and that one or two ships cannot possibly in the course of the next short summer, explore them all.

The Board of Admiralty, under the conviction of this fact, has been induced to offer a reward of £20,000 sterling to any ship or ships of any country, or to any exploring party whatever which shall render efficient assistance to the missing ships or their crews or to any portion of them.[6] This announcement, which even if the sum had been doubled or trebled, would have met with public approbation, comes however too late for our Whalers which unfortunately sailed before it was issued, and which, even if the ~~crews~~ news should overtake them at their fishing grounds, are totally unfitted for any prolonged adventure, having only a few months provision on board and no additional clothing.

To the American Whalers both in the Atlantic and Pacific, I look with more hope, as competitors for the prize, being well aware of their numbers and strength, their thorough equipment, and the bold spirit of adventure which animates their crews. But I venture to look even beyond these. I am not without hope that you will deem it not unworthy of a great and kindred nation to take up the cause of humanity which I plead in a national spirit, and thus generously make it your own.

I must here in gratitude adduce the example of the Imperial Russian Govt which (as I am led to hope by His Excellency the Russian Ambassador in London who forwarded a memorial on the subject) will send out exploring parties this summer from the Asiatic coast / of Behrings Straits northward in search of the missing vessels[.][7] It would be a noble spectacle to the world, if three great nations, possessed of the widest empires on the face of the globe, were thus to unite their efforts in the truly Christian work of saving their perishing fellow men from destruction. It is not for me to suggest the mode ~~by~~ in which such benevolent efforts might best be made. I will only say, however, that if the conception of my own mind to which I do not venture to give utterance, were realized, and that in the noble competition which followed, American seamen had the good fortune to wrest from us the glory (as might be the case) of solving the problem of the unfound passage, or the still greater glory of saving our adventurous navigators from a lingering fate which the mind sickens to dwell on, though I should in either case regret that it was not my own brave countrymen in those seas whose devotion was thus rewarded, yet should I rejoice that it was to America we owed our restored happiness, and should be for ever bound to her by ties of affectionate gratitude.

I am not without some misgivings while I thus address you. The intense anxieties of a wife and of a daughter may have led me to press too earnestly on your notice, the trial under which we are suffering[8] (yet not we only, but hundreds of others) and to presume too much on the sympathy which we are assured is felt beyond the limits of our own land. Yet if you deem this to ~~to~~ be the case, you will still find I am sure, even in that personal intensity of feeling, an excuse for the fearlessness with which I have thrown myself on your generosity, and will pardon the homage I thus pay to your own high character, and to that of the people over whom you have the distinction to preside.

I have the honor to be Sir,
with great respect, your obedient Servant
(Signed) Jane Franklin

5

To: LORDS OF THE ADMIRALTY
4 Spring Gardens, London, [Wednesday] 2 May 1849
Source: Handwritten transcript, SPRI, MS 248/212/1
Editorial comment: A transcription in Lady Franklin's hand. "1" is added in pencil at the top centre of the page, as well as "Copy" to the left-hand top (this in Lady Franklin's hand). [*sic*]: Investigators.

4. Spring Gardens
2[d] May. 1849

My Lords,

At a moment when my anxiety for the fate of my long absent husband and of his companions, forbids me to leave any means untried in their behalf, I claim respectfully your consideration of my present application.

You are already aware that the Whalers (to whom the Gov[t] reward of £20 000, offered for successful search would have been a great stimulation to exertion) had sailed before the pecuniary incentive to their exertions was made known, and that they are unprepared and unfitted ~~for~~ to avail themselves of it, even should the news reach them while yet at their fishing grounds.[1] It will not, I fear, be claimed by any of them, & the time which has elapsed since its announcement proves that no other parties have been induced by it to make the large outlay of capital which would be required for the adventure.

I ventured to suggest to the first Lord,[2] when the reward was first proposed, the expediency of adding £5000 to the £20 000 to give the latter effect, – I mean, to enable a vessel to start, a suggestion which was, I believe, strongly supported by others more competent than myself to judge of its importance; & I do not doubt that had time permitted, the consideration of this part of the question (for you kindly hastened your decision on account of the approaching departure of the American Mails) you would have brought it to a favorable issue.

I now venture again to submit this particular point to your consideration, though in a different form; since I now beg permission to ask that only a 4th part of the Reward itself should be granted at once for the fitting out of a vessel, leaving thus only 15000<£> reward to that particular vessel, in the event of her being successful.

If your Lordships do me the favour of granting this request, it will come in material aid of efforts I am myself making for the especial purpose of searching those parts which I believe to be unprovided for by any means within Sir James Ross's power & in which I have reason to believe the lost ships are as likely to be found as in any other, if not more so. / My anxiety is particularly directed to the passage known by the name of Wellington Strait or Channel, which has never yet been tried, and whose termination or Opening E. & W. are entirely unknown. – You are well aware that Sir John Franklin's attention was directed by your Instructions to this passage, in the second place, if he were baffled in his attempts to force his way in a S. W. direction from Cape Walker; and yet it is only slightly alluded to in the Instructions given to Sir James Ross when he sailed,[3] and is not mentioned at all in the Memorandum given by that Officer to Captain Bird for his guidance at a later period, when Sir J. Ross probably found that much that was intended to be done during the first season must be abandoned, he himself having to undertake a more westerly course.[4]

There are two other parts of scarcely inferior importance to Wellington Channel; one of them is Prince Regent's Inlet on Gulf of Boothia, which is now known through Dr Rae's Discoveries to have no Westerly outlet, but which in the Chart with which my husband was furnished, is laid down with a passage to the Westward.[5] The other is Jones' Sound and Smith's Sound W. & N. of Baffin's Bay, the former of which, especially, there is reason to believe, is connected with Wellington Channel. These parts my husband intended if possible to examine before his return home, if he had met with insurmountable obstacles elsewhere. It must be quite clear, I think, that if these be not examined, a very imperfect search, indeed, has been made for the lost Expedition.

Sir James Ross's personal investigation was directed to a totally different quarter from those above mentioned, and even in this quarter, the search is probably to begin only in the approaching summer, for as the Board of Admiralty is well aware, the last season was almost entirely lost by the lateness of his arrival at Lancaster Sound, owing, not to his setting off too late, but to the impediments from ice which he met with up to the latter end of August. Thus the Arctic Officers in Committee have recorded their unanimous opinion that the search for the missing Expedition has virtually to be commenced this year.[6]

It is then evidently a physical impossibility that these parts, including of course that about lat: 72 N. & long: 103–4 W. where Sir James Ross thought it probable the ships would be found, hampered in ice, with the coast of N Somerset to the E. should be searched during the short space of little more than 6 weeks which forms the Arctic Summer Season, especially as the / Investigators movements must be cramped by the orders she is under to watch the arrival of communication from England.

When you reflect upon these facts, you will I hope feel that I have only too much reason for my intense anxiety, and will not wonder that at this late moment, though it is not too late for my purpose, I turn to my own resources, as my last, but a necessary remedy. Had the storeship which you are about to send to the Arctic Sea been retained on the Station, as an additional means of search, during this & the following summer, I would not have troubled you with this application, and I beg you to give to my earnest and respectful request, no <other> interpretation than that which I am entitled to expect from you, viz: that it is a proof of my deep conviction that there is cause for great anxiety as to the fate of the missing expedition, now reduced to the last extremity, unless something more be done.[7]

I think I may use as an additional plea for your aid in the object I have in view, that every additional vessel sent out, adds greatly to the safety of the whole, and whatever inevitable risk attends all such expeditions is one which Volunteers alone are eligible to encounter – The Lords of the Admiralty will not admit of an objection which is refuted by their own liberality in offering a large reward for additional search, and their own anxiety I am persuaded must be, that whatever efforts are made, should be made as well as possible.

I have been informed that two such vessels as Dockyard Lighters with some alterations in the rigging, would be well adapted for my purpose, and being very strong, they would soon be made ready for sea.[8]

I should feel it to be a very great favour conferred on me, and an earnest of the success of my humble endeavour, if the Board would consent to grant me the loan of two such small vessels, or any other or others, fitted for the purpose, or permission to purchase them, if that can be done & my means admit of it.

I cannot attempt to conceal from the Board, that it is only by the sacrifice of all my private property (though I am not able to carry this to the full extent of what I desire) and by the additional aid of borrowed capital, that I shall be able to effect my object, if unassisted by them;[9] but I will still by the blessing of God carry it thro', certain that I shall never repent of my resolution, let what will happen, and feeling that it will be a real consolation to me, in the depth / of my own bereavement, if by this means, the humblest individuals only of that gallant band are saved from perishing.

When you humanely offered a large pecuniary reward for additional search, you could not have intended that I alone, the most interested perhaps of all others, in the success of this incentive to exertion, should be precluded from making any personal effort, because I am influenced by motives of a far different & higher character.

It will not be, I trust, and ought not to be a reproach to me, that I use every means & argument I can think of, that is upright and true, to move you to the consideration of my request, and if you ever have cause to look back upon any part of your administration with regret, which I hope may never be, it will not I am persuaded arise from your having extended to me the assistance I seek on the present occasion.

I have the honor to be
My Lords
Y^{r} obed^{t} Servant
(signed) Jane Franklin

6

To: JOHN CLAYTON[1]
[21] Bedford Place, London, [Thursday] 24 May 1849
Source: Printed transcript, *HCSP*, 1850/XXXV/327
Heading: THE Lady of Sir *John Franklin* to Mr. Clayton.

Sir, Bedford-place, London, 24 May 1849.

THE letter with which you have kindly honoured me, conveying the reply of the President of the United States to the appeal I ventured to address to him in behalf of the missing Arctic Expedition under my husband's command, has filled my heart with gratitude, and excites the liveliest feelings of admiration in all who have had an opportunity of seeing it.[2]

Relying upon the reports in the American papers just received, I learn that the people of the United States have responded, as you foresaw they would, to the appeal made to their humane and generous feelings, and that in a manner worthy of so great and powerful a nation, – indeed, with a munificence which is almost without a parallel.[3]

I will only add, that I fully and firmly rely upon the wisdom and efficiency of the measures undertaken by the American government.

I beg you to do me the favour of conveying to the President the expression of my deep respect and gratitude, and I trust you will accept yourself my heartfelt acknowledgments for the exceedingly kind and feeling manner in which you have conveyed to me his Excellency's sentiments.

I have, &c.
(signed) *Jane Franklin*.

7

To: BENJAMIN DISRAELI
4 Spring Gardens, London, [Wednesday] 6 June 1849
Source: Original, H, A/IV/N/3

4. Spring Gardens
6 June ./49

Dear Mr Disraeli

The remembrance of the very kind interest you took in a subject which deeply concerned me some time ago,[1] induces me to send you the inclosed letters.[2] I hope you will pardon me for thus intruding myself on your attention. I am indeed ashamed of it & would rather be kept out of sight as much as possible in the treatment of the question, but as it will be brought forward in the House tomorrow night by Sir Robert Inglis,[3] I think a few eloquent words from your lips would be deeply appreciated by the American people & I hope you will think they deserve such eulogy –[4]

Believe me dear Mr Disraeli
very truly yours Jane Franklin

8

To: BENJAMIN DISRAELI
Alverstoke Parish,[1] [Friday] 15 June 1849
Source: Original, H, A/IV/N/4
Editorial comment: [*sic*]: yesterday. Lady Franklin refers to the *Times* of 13 June, not 14 June.

Alverstoke 15 June
1849

Dear Mr Disraeli

I read in the Times yesterday your kind support of the tribute of commendation to the American government & beg to assure you of my sense of the obligation you have conferred on me.[2]

In the same paper I was thrilled & vexed to see my own letter in print ushered into notice by Mr Jesse.[3] I have not the pleasure of knowing that gentleman, but have no doubt that he must have entirely misunderstood the intentions of the kind friend who placed those papers / in his hands. –[4]

I am, dear Mr Disraeli
very truly yours
Jane Franklin

9

To: ZACHARY TAYLOR
[4] Spring Gardens, London, [Tuesday] 11 December 1849
Source: Printed transcript, *HCSP*, 1850/XXXV/330–1
Heading: The Lady of Sir *John Franklin* to the President.
Editorial comment: This letter belongs in a larger collection of *HCSP*, given to Parliament by Taylor and serving to explain his inability to organize an expedition in 1849, and his 1850 determination to.

Sir, Spring-gardens, London, 11 December 1849.

I HAD the honour of addressing myself to you, in the month of April last, in behalf of my husband, Sir John Franklin, his officers and crews, who were sent by Her Majesty's government, in the spring of 1845, on a maritime expedition for the discovery of the North-west Passage, and who have never since been heard of.

Their mysterious fate has excited, I believe, the deepest interest throughout the civilized world, but nowhere more so, not even in England itself, than in the United States of America. It was under a deep conviction of this fact, and with the humble hope that an appeal to those generous sentiments would never be made altogether in vain, that I ventured to lay before you the necessities of that critical period, and to ask you to take up the cause of humanity which I pleaded, and generously make it your own.

How nobly you, sir, and the American people, responded to that appeal, how kindly and courteously that response was conveyed to me, is known wherever our common language is spoken or understood; and though difficulties, which were mainly owing to the advanced state of the season, presented themselves after your official announcement had been made known to our Government, and prevented the immediate execution of your intention, yet the generous pledge you had given was not altogether withdrawn, and hope still remained to me that, should the necessity for renewed measures continue to exist, I might look again across the waters for the needed succour.[1]

A period has now, alas! arrived, when our dearest hopes as to the safe return of the discovery ships this autumn are finally crushed by the unexpected, though forced return of Sir James Ross, without any tidings of them, and also by the close of the arctic season.[2] And not only have no tidings been brought of / their safety or of their fate, but even the very traces of their course have yet to be discovered; for such was the concurrence of unfortunate and unusual circumstances attending the efforts of the brave and able officer alluded to, that he was not able to reach those points where indications of the course of the discovery ships would most probably be found.[3] And thus, at the close of a second season since the departure of the recent expedition of search, we remain in nearly the same state of ignorance respecting the missing expedition as at the moment of its starting from our shores. And in the meantime our brave countrymen, whether clinging still to their ships, or dispersed in various directions, have entered upon a fifth winter in those dark and dreary solitudes, with exhausted means of sustenance, while yet their expected succour comes not!

It is in the time, then, of their greatest peril, in the day of their extremest need, that I venture, encouraged by your former kindness, to look to you again for some active efforts which may come in aid of those of my own country, and add to the means of search. Her Majesty's Ministers have already resolved on sending an expedition to Behring's Straits,[4] and doubtless have other necessary measures in contemplation, supported as they are, in every means that can be devised for this humane purpose, by the sympathies of the nation and by the generous solicitude which our Queen is known to feel in the fate of her brave people imperilled in their country's service. But, whatever be the measures contemplated by the Admiralty, they cannot be such as will leave no room or necessity for more, since it is only by the multiplication of means, and those vigorous and instant ones, that we can hope, at this last stage, and in this last hour, perhaps, of the lost navigators' existence, to snatch them from a dreary grave. And surely, till the shores and seas of those frozen regions have been swept in all directions, or until some memorial be found to attest their fate, neither England, who sent them out, or even America, on whose shores they have been launched in a cause which has interested the world for centuries, will deem the question at rest.

May it please God so to move the hearts and wills of a great and kindred people, and of their chosen Chief Magistrate, that they may join heart and hand in the generous enterprise! The respect and admiration of the world, which watches with growing interest every movement of your great republic, will follow the chivalric and humane endeavour, and the blessing of them who were ready to perish shall come upon you!

I have, &c.

(signed) *Jane Franklin.*

His Excellency the President of the United States.[5]

10

To: JOHN CLAYTON
[33] Spring Gardens, London, [Wednesday] 12 December 1849
Source: Printed transcript, *HCSP*, 1850/XXXV/331
Heading: The Lady of Sir *John Franklin* to Mr. *Clayton*.

Sir, Spring-gardens, London, 12 December 1849.

I BEG to transmit, through you, to the President of the United States, the accompanying letter, trusting to that same kindness on his part of which you have already been the feeling and eloquent exponent, that it will be received with an indulgence similar to that which I met with before.

Sir John Richardson informs me that the interest felt in the United States for the unfortunate missing expedition under my husband's command is as lively and deep as ever; and in this feeling, and in the benevolent disposition of the President, I place my humble confidence.[1]

I have, &c.
(signed) *Jane Franklin*.

11

To: [WILLIAM HAMILTON][1]
33 Spring Gardens,[2] London, [Wednesday] 23 January 1850
Source: Printed transcript, *HCSP*, 1850/XXXV/316
Heading: No. 14 (B.) | COPY of a LETTER from Lady *Franklin to the Secretary of the Admiralty.*
Editorial comment: This letter is one of a collection of similar letters sent by Lady Franklin to the Admiralty on behalf of private individuals interested in participating in the search. It has been selected for publication because of the significance of the people mentioned (see endnotes for Introduction).

Sir, 33, Spring-gardens, 23 January 1850.

I BEG you to do me the favour of laying before the Lords Commissioners of the Admiralty a letter which I have received from Mr. Robert Goodsir, late surgeon of the "Advice" whaling-ship, Capt. Penny, and brother of the assistant-surgeon of the "Erebus," and I must express my earnest hope that their Lordships will be pleased to enable me to afford that encouragement to his request for which he looks with so much earnestness.[3]

In connexion with this subject, I take the liberty of adding, that I have within the last two or three days received a letter from Captain Penny, expressing with much anxiety his fears that any farther delay in the preparations for the renewed search from the east must be fatal to its success.[4]

I had the pleasure of receiving, yesterday, through the kindness of the Lords Commissioners, the copy of an abstract made by Captain Kellett of his voyage, for which I beg to return my best acknowledgements.[5]

I am, &c.
(signed) *Jane Franklin.*

12

To: BENJAMIN DISRAELI
33 Spring Gardens, London, [Wednesday] 30 January 1850
Source: Original, H, A/IV/N/6
Editorial comment: On the fourth page of the manuscript, Mary Anne Disraeli has written "1850. Jany 30th | Lady Franklin."

33 Spring Gardens
30th Jany 1850

Dear Mr Disraeli

The Admiralty have not thought fit to take upon themselves the responsibility of renewing the search for the missing Arctic Expedition in the quarter where it has been interrupted by Sir James Ross's premature return, without the sanction of Parliament. They expect an expression of public opinion on this point, and we trust that all parties may be enabled to unite in that expression.

It is not I hope presuming too much when I venture to believe that the same generous support which you have given us on former occasions will not be denied us at a crisis when life ~~and~~ or death seem to hang on the decision.

We / ask only that the search which Sir J. Ross was instructed to make and which was intended, if necessary, to last for 3 years, may not be abandoned because a concurrence of untoward & very unusual circumstances drove him back in little more than half that period, without his having visited a single spot where the missing navigators or their traces were most likely to be found. – Delay at this moment is fatal to us – there has never yet sailed an exped^{n} which was not in preparation at the end of Jan^{ry} whereas this which we hope for, is not yet <even> begun.[1]

Excuse me for occupying the smallest portion of your time at a moment when such important cares & duties rest upon you, & believe me

dear Mr Disraeli
yours truly & obliged
Jane Franklin

I send you, without expecting you to read it, a book on the present state of the Arctic question by Dr Scoresby himself a veteran Arctic navigator[.][2] It contains the Admiralty charts.

13

To: [WILLIAM HAMILTON]
Bedford Place, Brighton, [Monday] 31 January 1853
Source: Handwritten transcript, SPRI, MS 248/212/5
Editorial comment: "To the Sec of the Admiralty / London" is written in pencil at the top of the page, as well as "5." The letter is in Sophia Cracroft's hand, while the postscript is in Lady Franklin's.

Bedford Place
Jany 31/53

Sir,

I beg to express to you my thanks for the courteous terms in which the reply of the Board of Adm^{ty} to my letter of the 13th inst. was conveyed to me, & to inform you that, deeply regretting the result of my application, I feel myself under the painful necessity of endeavouring to supply in some measure by my own humble exertions that deficiency in the Gov^{t} measures to which I presumed to call their Lordships' attention.[1] To do otherwise, so long as I have any pecuniary means at my disposal, or any hope of moving my countrymen to sympathy with the object in view would be a dereliction of duty on my part to which no counteracting consideration would ever reconcile me. I have therefore only to request respectfully that you will do me the favor of allowing the stores & fittings lately removed from the "Isabel" screw steamer now in Woolwich Dockyards, to be replaced in her in order that she may be enabled to proceed upon the voyage to Behring Strait with as little delay as possible.[2] I need scarcely explain that the Isabel steamer was made over by me last summer equipped & manned for Captain Inglefield as his own property, / on condition of his employing her during the remainder of that season in Arctic search, and he now kindly restores her to me in this emergency –, I, of course, paying all the expenses of his equipment, outfit & voyage –

As you have kindly assured me that I shall not meet with less consideration & assistance from the present Board of Admiralty than was accorded to me by the late or any preceding Board, I am induced to trust that there will be no objection made to my request.[3]

I will add that a considerable portion of the funds by which I hope to carry out this Expedition will be generously supplied to me by the colony of V. D. L. (of which my husband was at one time Governor)[.] It is sent to me, quite unforeseen on my part, as a testimonial of attachment & respect to him, & of sympathy with the private efforts made in his behalf and her especial reference to the expedition of the "Isabel" in Behring St where the Colonial subscribers believe that vessel to be now engaged in the search. I am promised assistance from other quarters also.

Trusting that you will be pleased to give me a favorable and early reply – I have the honour to be, Sir,

your Obedt Sert
Jane Franklin.

\+ the colony sent £1700. /

P.S. Since writing the above, I have received the enclosed private letter from the Secretary of the U.S.N. and a chart which he begs me to present to the Board of Admiralty.[4]

I believe you may find Mr Kennedy's letter not devoid of interest, & indeed the postscript seems to require that you should see it. It is in reply to a letter of thanks which at the particular request of Dr Kane I addressed to the Secretary in acknowledgement of the assistance promised to that officer in his private expedition.[5] I had no knowledge, however, of that Exptn till it was communicated to me & decided upon by Dr. Kane, a fact I think it right to explain since the Secretary, with the kind desire no doubt of farther recommending it to the public & Congress has represented it ~~to~~ in another light – in his official Report to the President.[6] I have every reason to be extremely grateful to the American government – you will not I respectfully hope allow me to feel less so to you –.

14

To: SIR JAMES GRAHAM[1]
39 Bedford Square, Brighton, [Friday] 20 January 1854
Source: Handwritten transcript, SPRI, MS 248/212/6
Editorial comment: Letter is transcribed in an unknown hand. [*sic*]: 13th (Graham's letter, dated the 12th, is postmarked 14 January 1854); Lordships.

Copy

39 Bedford Square
Brighton
20th Janry 1854.

Sir

I have to apologize that it has not been in my power sooner to acknowledge your letter of the 13th but the shock it inflicted on my already shattered health was such as not to permit me any mental exertion until now.[2]

I beg to assure you of my sense of the kind expressions you have made use of in communicating to me intelligence which no amount of courtesy, however, could render otherwise than most startling and painful – not the less so, because the Board of Admiralty having so recently repudiated all participation in the authoritative declaration of the "Times" which first announced their intended measure, I was not prepared for the proof of that paper having been really the authentic exponent of their intentions.[3]

It was not so much, however, the assurance thus given me on their Lordships part, which relieved my apprehensions, as the conviction that it was impossible the Admiralty could contemplate such a step as the withdrawal from their list of the living, of the officers of the Erebus and Terror, while the search for those officers and their crews was actually being carried on under their own directions, and which the result of such search could not by any possibility be known before the autumn of the present year.

It may also be observed that your Expedition under Sir Edward Belcher is so far from being prolonged beyond the period contemplated when it was fitted out, that it has not yet completed its anticipated term of service, / and accordingly, only 6 months ago, it was furnished with fresh supplies for continuing the search.[4]

Another consideration confirmed my incredulity as to the authenticity of an announcement which filled almost every heart with astonishment and dismay. It has been the careful and emphatic declaration of every successive Board of Admiralty since the search for the missing ships began, that no motives but those of the purest humanity influenced their proceedings, and though my husband was sent out to discover a N. W. Passage, yet no such purpose was ~~to be~~ admitted to be the aim and end of those who went to rescue their – imperilled and suffering countrymen – Could I have expected then, that you, Sir, and your Colleagues, would have taken the earliest opportunity of shewing that as soon as the N. W. Passage was discovered, all farther interest in the search was at an end? I have said the earliest opportunity, but it is without an opportunity such as the return of Sir Edward Belcher might have afforded, that you have resolved on this premature and cruel measure![5]

No act of man's expediency however, justified as the object of it may be by consideration of economy, and by critical political circumstances, can put an end to the breath of life at the end of the financial year. No Gazette of such a nature can bear the impress of truth upon it.

It is indeed greatly to be ~~[illegible]~~ <deplored> that the desired stoppage of payment cannot be effected (if it cannot) without a resort to this deplorable and revolting assumption – I believe that the positive declaration you announce to me of the death of 135 men, on or by the 31st of next March to be presumptuous in the sight / of God, as it will be felt to be indecorous, not to say indecent (you must pardon me for speaking the truth as I feel it) in the eyes of men.[6]

Yet I would not be misunderstood. That very great and painful certainty exists upon the subject, no one can deny, nor am I insensible to the reasonableness of putting a stop to the expedition before that uncertainty ceases, since if nothing farther be done to throw light upon the mystery, it may continue to the extent of the natural lives of the persons concerned.

You will not expect from me, Sir, to claim from the Admiralty the Widow's pension which you remind me is granted under certain regulations.[7] – I believe that my husband may yet be living where your expeditions have never looked for him, in a quarter where my own little vessel, in the absence of any better means, would have endeavoured to look for him had you not denied me the means and facilities of doing so.[8]

These are simple facts which it requires no Arctic authority to judge of, tho' I believe there would be few amongst the Arctic officers to deny that my husband went up Wellington Channel, and there are none who can dispute that between Wellington Channel and Behring St. there is a space through which his course must have been taken which has never yet been explored. If I have any right to speak more confidently than another upon this subject, it is because I am in possession of my husband's mind upon it, and that of many of those who sailed with him.[9]

It is not my intention at this moment to trouble you with any observations on the legal view you appear to have taken of the question.[10]

Subsequently to the letter with which you favored / me, I received a communication from the Board to which I shall reply as soon as my health permits, giving the grounds on which I must respectfully remonstrate against the measure announced to me, in order that my remonstrance may remain on record.[11] In the meantime as I have no hope of being well enough to write immediately, I shall feel obliged by your communicating to them this letter.

I have the honor to be Sir
yr. obed^{t} Serv^{nt}
(signed) J. F.

To The Right Hn^{ble}
Sir James Graham &c.

15

To: [WILLIAM HAMILTON]
[4] Spring Gardens, London, [Thursday] 9 February 1854
Source: Original, PRO, TS 18/413/1
Editorial comment: [*sic*]: a missing closing parenthesis. This letter is in a box with a series of correspondence regarding the Admiralty *Gazette* notice of 19 January, including notes from Lady Franklin, the Office of the Inspector of Seamen's Wills, the accountant general of the navy, and Sir James Graham.

Spring Gardens
Feby 9th 1854

Sir

I have to request that you will do me the favor of moving my Lords Commissioners of the Admiralty to grant me the sum of £1000 from the arrears of my husband's pay, or such further portion of it, as, together with the sums I have already drawn, will amount to one half of the arrears due to him since his departure from England up to the 31st of March next.[1]

I make the request as needing this sum for subsistence, as well as for the liquidation of debts incurred in my efforts to save my husband's life, and because he authorized me to draw his pay if I required it.

I have refrained hitherto from making any regular periodical application for this money as a legitimate part of my income (in the / manner in which all the other officers' wives of the –Erebus" & –Terror" have I believe done, mainly from my repugnance to present such repeated requests when I had so many others to urge upon their lordships connected with my private expeditions in which the Admiralty alone could help me.

I have the honor to be
Sir
your obed[t] serv[t]
Jane Franklin

to the Secretary
of the Admiralty

16

To: LORDS OF THE ADMIRALTY
4 Spring Gardens, London, [Friday] 24 February 1854
Source: Printed transcript, *HCSP*, 1854/XLII/339–50. Three rough drafts of this letter, in both Lady Franklin's and Sophia Cracroft's hands, are extant in SPRI, MS 248/212/7.
Heading: ARCTIC EXPEDITIONS. "COPY of LETTER addressed by Lady *Franklin* to the LORDS COMMISSIONERS of the ADMIRALTY, dated the 24th day of February 1854, in reference to their Lordships' Announcement in the London Gazette of the 20th day of January 1854, respecting the Officers and Crews of Her Majesty's Ships 'Erebus' and 'Terror.'"
Editorial comment: This is the most widely circulated – and by far the longest – of Lady Franklin's letters to the Admiralty. It was published without her consent and was printed in pamphlet form and generally available (see Introduction). [*sic*]: Blenby; predicated. HMS *Enterprise* has inconsistent spelling in private and House of Commons documents, and so is not acknowledged here. In the drafts of this letter (SPRI), and particularly when Sophia Cracroft transcribes the polished version, "Blanky" looks remarkably like "Blenby." This error may in fact be Sophia Cracroft's, and not the Admiralty typesetter's.

My Lords, 4, Spring Gardens, 24 February 1854.

In a letter which I had the honour to address to the First Lord of the Admiralty on the 20th of January, and which at my request he kindly forwarded to the Board, I expressed in language of deep emotion the feelings of pain and wonder to which your summary and unexpected sentence on my husband, Rear-Admiral Sir John Franklin, and the officers and crews of the "Erebus" and "Terror," had given rise.[1]

You have been pleased to explain to me, in terms of which I cannot but acknowledge the courtesy, that it is for the convenience of winding up the accounts of the financial year, which closes on the 31st of March, that you have fixed upon that moment for consigning 135[2] seamen in Her Majesty's service simultaneously to the grave, unmindful of the discordant fact that Her Majesty's ships on the other side of the Atlantic are now, and will be on 31st of March next, preparing to discover the abodes of these very men, considered as living beings, yet to be rescued.[3]

My Lords, I make no vain complaints of the manner in which your Lordships' intentions have been communicated to the public, distressing as it is to the feelings of the living, and little respectful as it has the semblance of being to the memory of those who, if they have "died in Her Majesty's service," / might have been deemed entitled to more regretful mention. All who are most deeply concerned in this announcement must be well aware that nothing could be further from your Lordships' intentions than to produce such an impression; and we lose our painful sense of the hard official language of your Gazette notice in the severer shock which its meaning gives to those hopes and that reliance which we have hitherto placed in you, as, under God, our sole help and refuge.[4]

Neither, perhaps, can we presume to complain that an expenditure, which cannot be proved to be lawfully due, should be suspended, even had there been no immediate exigencies of the public service, if such exist, to justify its withdrawal. I believe there are few among the representatives of the absent who have not felt that the Admiralty have acted liberally, kindly and generously, in continuing, during years of uncertainty, the pay and wages, as if certain of their returning to claim their own.[5] The search might have gone on though the payment was suspended, and none would have doubted that on the safe return, however distant, of the rightful claimants, those wages, so hardly won, would have been paid them to the full, and their right standing in Her Majesty's Navy restored to them, even though other brave men had been worthily promoted to fill their vacant places. It is not, then, of the retrenchment, but of the reason on which you have thought fit to base it, that we have cause to complain.

Your Lordships say, in your Gazette notice, that the officers and crews of the "Erebus" and "Terror" are, on the 31st of March next, to be considered dead, if no intelligence arrive in the meantime to the contrary, your Lordships being aware that the arrival of any intelligence before that date is physically impossible.[6]

We knew, my Lords, that this sentence cannot realise the doom of its victims, whose possible return you are compelled for your own financial security to admit; that it is null as touching the fact can be considered no evidence in a court of law, and leaves the truth, whatever it be, untouched.

Yet does it sound on the public ear, and more deeply in the ear of many heart-anxious listeners, as the knell of departed hopes, the warning voice that tells us we are to prepare for the abandonment of those unhappy men to their fate. And if it be not so, and that your Lordships have used this language only as a means of legalising your financial measure, would that you had explained to us that the search now carrying on would not be affected by it, but would be continued till its especial object was accomplished according to the expectations raised reasonably and inevitably by your Lordships' own previous course of action. The special object of the present expedition was to search for the missing ships in that quarter of the Arctic seas where they had not yet been looked for; it was recommended by a great majority of Arctic officers appointed to consider the question, who believed that my husband and his companions had passed that way, and were yet to be found alive.[7]

The expedition of Sir Edward Belcher, founded on these conclusions, was provisioned for a certain absence of three years, and only six months ago was reprovisioned for a longer period. It is not yet two years since the expedition sailed, and it has not yet accomplished its mission nor been absent its expected term of service, nor can we obtain any information as to its proceedings till next autumn, nor perhaps then, unless a special messenger be sent for the purpose, nor shall we learn, perhaps, at that period, the total result of the explorations made or yet making.

These facts, so inaccordant with your Lordships' sentence of death, are the grounds of my hopes that that decree may not involve the fatal conclusion as to your intentions, to which, by a too inexorable reasoning, it would seem to lead. Yet in the meantime an unauthorised impression is produced, most discouraging and painful, tending directly to extinguish hope, to paralyse exertion, and even to suppress the expression of honest sentiment.

I am under the necessity, in spite of my innate trust in your Lordships' justice and compassion, of dealing with the "Gazette" notice as I found it, in its literal sense, and it must therefore be my endeavour to prove in this letter why I cannot accept your Lordships' sentence, but am compelled to record my respectful, but most earnest remonstrance and protest against it. /

The grounds on which it appears to me reasonable that my husband and his companions in the "Erebus" and "Terror," should not be considered dead, but living, are these:

First. Because no evidence has been discovered of any catastrophe having befallen them.

Secondly. Because the quarter of the Arctic Sea, where it is most probable that the missing parties would be found living, or their fate ascertained, has never yet, so far as we know, been explored; Sir Edward Belcher, when last heard of, having advanced only to the verge of the open sea to the North-west, but without entering it; and because the part thus indicated is one of the two courses pointed out to my husband in the Admiralty Instructions, for him to follow, and also because it has been pronounced, after a thorough examination of the other course, that he could not have passed that way.[8]

Thirdly. Because within this unexamined region the resources for supporting life are probably abundant;[9] and

Fourthly. Because my husband and his officers steadily contemplated, and from the first provided for a detention extending over an indefinite period, should difficulties occur to prevent their return at the time expected.

I. And first, as to the absence of all signs of wreck or disaster. This negative evidence of the safety of the expedition has been gained in every part of the Arctic Sea which has yet been visited. Neither the bodies of men, nor parts of ships, timbers, spars, stores of any description have been found, either afloat in the currents, or washed upon the shores.[10] In Wellington Channel, where the missing ships are known to have been, nothing has been found (beyond the signs of their well-being at winter quarters) but some drift pine wood, belonging to the forests of a milder climate to the North or West.[11]

The captains of the whaling ships, men the most experienced in such matters, concur in asserting that it is next to impossible that two ships, like the "Erebus" and "Terror," could be crushed and destroyed, without any of their crews escaping, and without some traces of the disaster being found; and one of our most distinguished Arctic navigators has very recently declared that he was never more strongly of opinion than he is now, that it is utterly improbable that Franklin's ships, men and all, have been destroyed by any accident among the ice. I do not think it necessary to adduce any contrary opinions, because they appear to have been adopted rather as a last resort for the solution of a so-called mystery, than from any indisputable data.

There was a time, it is true, when it was somewhat unscrupulously asserted that both the ships had been swallowed up in the ice in their passage across Baffin's Bay, during their first summer, and this opinion, which was utterly devoid of even the semblance of justification, obtained some degree of credit till the discovery of their first winter quarters, on the other side of the supposed field of disaster, put a summary end to the gratuitous tale. And next we were assured that our brave navigators, whose high sense of duty had never been questioned before, had deliberately turned their backs upon the work before them, after only one winter's absence, and been crushed, or had foundered on their way home.

It would be presuming too much on your Lordships' patience, to dwell on other absurd stories of murders, burnings, &c., invented by the mendacious half-caste Esquimaux, Adam Beck, when he desired to put an end at once to the search, in order to get earlier back to the home he had been enticed to leave.[12]

But there is yet a more recent report, which, visionary as I am myself disposed to deem it, on the authority of persons experienced in Arctic visual phenomena, has been deemed otherwise than necessarily a delusive appearance by persons entitled to every respect. And yet I need not argue in addressing your Lordships against this spectacle of the supposed "Erebus" and "Terror" drifting away on the top of an iceberg from some unknown quarter to the banks of

Newfoundland, since assuredly had your Lordships believed it at the time the report reached England, which was in the spring or summer of 1852, you would not have lost a moment in taking steps to search the shores which those ships, if / such they were, must have quitted, and where they must have left their human freight, still living, behind, since the spectators of the phenomenon affirm their conviction that there was not a living soul on board.[13] And I am the more persuaded of this, since it was at this very period that I offered for the Admiralty's acceptance, nay, entreated their acceptance of my little vessel, the "Isabel," equipped and provisioned for Arctic service, which had fallen into my hands after the failure of Mr. Beatson's expedition, and thus a search of the shores alluded to, at no further expense to Her Majesty's Government than the manning and officering of the little vessel, might have been effected without a moment's delay.[14] I could not myself, however, have been expected to submit to the particular attention of the Board a search after those iceberg ships, in which I had not faith, especially in the quarter to which it would probably have been directed, namely, the coast of Labrador. Nevertheless, if it should now be your Lordships' pleasure to cause inquiries to be made by any of your returning ships on the coast of Labrador, for the unfortunate people alive so late as 1851, and thus, as it is supposed, bereft of their floating home and means of transport, I could not but regard the measure with grateful satisfaction, though I may humbly express my opinion that it is not from the coast of Labrador that these supposed discovery ships could have drifted. But it is always an advance towards the undisputed settlement of our missing navigators' position, to know where they are not, and, indeed, I would readily give heed to this, or any other not impossible conjecture which promotes search, if it were not that by so doing, efforts are diverted from the only course which I believe to be the right one.[15]

But before closing these observations on the absence of all evidence of any fatal catastrophe having happened to the missing expedition, I am reminded of a passage in a despatch of Captain M'Clure, deposited on Melville Island, which has been exultingly quoted by a writer in the "Times," in confirmation of his opinions to the contrary.[16]

Captain M'Clure says, "It is my intention, if possible, to return to England this season, touching at Melville Island and Port Leopold, but should we not be again heard of, in all probability we shall have been carried into the Polar Pack, or to the westward of Melville Island, in either of which to attempt to send succour would only be to increase the evil, as any ship that enters the Polar Pack must be inevitably crushed." And again he says, "A ship stands no chance of getting to the westward by entering the Polar Sea, the water alongshore being very narrow and wind contrary, and the Pack impenetrable."

The value of these remarks of Captain M'Clure is, I conceive, to be limited by his personal observation and experience. It is evident he was speaking of that portion of the Polar Sea with which he was himself acquainted, without noting the distinction which appears to have been recognised of late years between the sea lying within 15 degrees around the Pole, and that section of it to the southward, between the chain of the Parry islands and the face of the American continent, which, in the earlier annals of Arctic discovery, was included in the general name of Polar Sea, and is so called in the published narratives of my husband's overland expeditions.

We have no description of ice seen in those northern seas which I believe the missing ships to have entered, corresponding to that encountered by Captain M'Clure in the narrow channels, and in the ice-clogged shores of Banks' Land and Melville Island, where it is supposed to be caused by the prolongation of the land westward towards Behring Strait. Dr. Scoresby has justly observed, that had the chain terminated near the meridian of Mercy Bay, a far wider space of open water should have been observed there, after the southerly gale, than seems to have occurred.[17] It appears, therefore, not reasonable to draw any unfavourable conclusion as to the safety of ships which entered a Polar Sea north of the chain of islands, from any appearances which come under the observation of Captain M'Clure in his lower latitude and confined position. And this reasoning, as it affects the safety of the "Erebus" and "Terror," will be the more readily admitted when it is remembered that the passage between Melville Island and Banks' Land, respecting which Captain M'Clure's observation was made, was expressly pointed out to Sir John Franklin, in his instructions, to be *avoided*. It was the only part of the Arctic seas which he was enjoined not to approach, and whatever therefore may be the nature of the / ice within or near, the missing expedition has assuredly avoided exposure to it. For this reason I have always been persuaded that there was no probability of the ships ever being found, or even retreating upon Winter Harbour, or anywhere on the south coast of Melville Island, which has been the object of so many attempts both from the east and from the west, and has become, in fact, in consequence of its halfway position, the rendezvous of the searching squadron.

I may add, that though in that portion of the Polar Sea which Captain M'Clure had in his eye, he believes no ships could live, yet that towards that other part of the Polar Sea, which I doubt not my husband entered, a little solitary vessel of less than 150 tons, bearing the American flag, is now dauntlessly pursuing her way, undeterred by any conjectural dangers, but aiming to solve in some degree, in subordination to the higher object of humanity, the geographical problem of what exists further north than any discoverer has yet penetrated.[18]

II. I have ventured to make the assertion that my husband and his companions have never yet been looked for in that part of the Arctic seas where a great probability exists that they would be found.

By your Lordships, who are acquainted with the proceedings of every successive expedition, of each division of the search, and of the results of the whole, this position will hardly be disputed. To the great majority of the public who have heard of one costly expedition after another, and believe that by this time the superficial area of Arctic waters must well nigh be swept, it would appear a startling assertion difficult of acceptance.[19] Yet it requires but a transient glance at the Polar chart, as it appears with the very latest geographical acquisitions, to see that between the meridian of Wellington Channel and Behring Strait, there lies a blank space, in which neither sea, nor coast, nor island is laid down. It comprises about 70 degrees of longitude, or, if measured, in the parallel of 78 degrees (the most southerly which could apparently have been navigated by the discovery ships), between 800 and 900 miles. What discoveries my husband may have made in this space he has not returned to tell; no one has followed him there. And lest it should be objected that I assume too much in asserting that this must have been the course taken by the missing ships, allow me a few further explanations.

It cannot be denied that the Admiralty Instructions presented to my husband two routes by which to endeavour to effect a passage from the Atlantic to the Pacific. The southern route, by Barrow Strait to the south-west, was recommended to him in preference, or in the first instance, and was the object of his own predilections, inasmuch as, if practicable, it would lead to the northern front of the American Continent, already known to him by his own and other surveys. But my husband anticipated that before arriving on these familiar shores, he would probably meet, in particular latitudes and longitudes which he pointed out, with insuperable obstacles, and accordingly his mind was greatly occupied before his departure with the alternative course, or northern passage, which was wholly new, and moreover extremely popular amongst his officers. It was anticipated with enthusiasm by Captain Fitzjames, the commander of the "Erebus," that the ships would descend from a high northern latitude upon the coast of Asia, and that he would be sent home with my husband's despatches through Siberia.[20]

One thing I can affirm with certainty, that it was my husband's determination to try both routes before his return,[21] and so strong was the feeling of interest and importance attached, as it was said, by the Admiralty themselves to that which led into the unknown northern waters, that a confident expectation existed that even should the ships effect a passage into the Pacific by the southern course, they would, on their return to England, be dispatched afresh on another voyage of discovery to report upon the high Polar regions.

In confirmation of these facts, or of the inference to be derived from them, we find the first winter quarters of the "Erebus" and "Terror" at the opening of Wellington Channel, where the northern route separates from the western one, and whence each could be watched with advantage. The passage westward was probably first tried, and found impracticable, as it afterwards proved / to

Sir James Ross and Captain Austin, and thus with increased eagerness and solicitude would the commanders of the expedition fix their attention upon the northern passage.[22] It has been found that their sledge tracks were multiplied in this direction, and at a lookout station commanding it, papers were picked up, showing that the watches were unremitting. Who can doubt that the same open water, seen by Captain Penny from the heights of Cape Spencer,[23] verified afterwards by himself and his parties, when so early as the month of May the progress of the sledges was arrested by it, – who can doubt that this open water was first seen by the observers of the "Erebus" and "Terror," and that the earliest disruption of the intervening barrier of ice was the signal for their departure from Beechey Island, accomplished, as it evidently was, in haste, but without disorder? All the conjectural difficulties and impediments which were brought before the Arctic Committee contrary to this presumption, have been overthrown by the undeniable fact that Sir Edward Belcher has since carried his ships in clear water up the same channel, even to the very verge of an expanse of sea to the north-west, to which he saw no limit, and which is all we require to complete the presumptive evidence.

On the negative evidence, in favour of my husband's having taken the northern passage, a few words will suffice. It is unnecessary to enter minutely into the researches of each expedition, especially as all, or almost all, had to go over the same ground before they could make a step in advance; and as this ground was minutely examined by each party in succession, we had the discouraging report of "no traces of Sir John Franklin" echoed and re-echoed, till it produced, I believe, upon the minds of many who were not aware of this explanatory fact, the painful and delusive impression that it was of no use looking any longer for those who were not anywhere to be found.

The coasting expedition of Sir John Richardson between the Rivers Coppermine and Mackenzie, first proved that the ships he was seeking had not arrived on that part of the American coast, and consequently, as it might be inferred, on the coast further to the westward.[24] Thus the examination of this closing portion of the south-western route narrowed the search, but it could not prove that my husband had not taken that course, and been arrested in it by obstacles in the nearer and earlier portion of it. This conclusion, however, was at length obtained by a branch of Captain Austin's expedition in 1851, when Captain Ommanney and Lieutenant Osborn, advancing beyond the limits of previous explorers, examined on foot the coast which tends southward beyond the 100th degree of west longitude, being the precise course pointed out to my husband in his Instructions. These two officers came to the conclusion, not only that he / had not passed that way, but that there was no navigable channel for ships in that direction. There remained no other reasonable alternative but that the missing ships had passed up Wellington Channel, which had always been little less probable than the other course, though as yet unexamined.

If my own steadfast convictions as to my husband's plan of action can give any additional force to the arguments I have already adduced, my Arctic friends will bear witness to the anxiety I have ever felt for the exploration of the route through and beyond Wellington Channel from the earliest period. It was not, however, till the intelligence received from Sir James Ross, at the close of 1848, showed the improbability of its coming within the sphere of his operations, that I ventured to implore the Admiralty to make it a distinct and primary object of their attention. Failing to persuade them to allow the "North Star," then about to be dispatched with supplies for Sir James Ross, to remain out for the examination of this strait, and failing also in my endeavours to equip a private vessel for the express purpose of examining at least the headlands of the channel, I went to Dundee, and asked Captain Penny if he would undertake the search, in case that the Admiralty could be induced to accept his services.[25] His enthusiastic reply was followed (after the return of the Government expedition the same autumn, 1849) by his submitting his wishes and plans to the Board of Admiralty, and grateful indeed did I feel for their ultimate acceptance of his proposals; and when Captain Austin's nobly-appointed expedition was organised, it appeared to me that both the south-western and the northern routes were now sure to be thoroughly explored.

It was not till then that I dispatched my own little vessel, the "Prince Albert," into Regent Inlet (which was not included in the instructions given to either expedition), under the idea that if the crews of the "Erebus" and "Terror" had been forced to abandon their ships on the south-western route, for we were yet ignorant which route they had taken, they might retreat towards the perhaps nearer and well-known resources in that quarter, rather than upon the more distant and more barren shores of North America.

The "Prince Albert," in her way back, touched at Beechey Island, and brought home, in September 1850, some small vestiges of the missing ships, which had been found at Cape Riley, and reported the traces of an encampment.[26] A year elapsed, and then came accumulated and exuberant evidence of the winter quarters of the ships, and their prosperous condition at this spot up to the spring of 1846.[27]

But it was the searching ships themselves that brought home prematurely the news. The south-western course had been explored in vain, but this northern one had not been attempted. Doubt even was thrown on the open water expanding to the north, which had been seen by Captain Penny, that most important discovery which alone, at this critical period, saved the pioneers in the "Erebus" and "Terror" from being consigned to destruction. But their doom was delayed.

In 1852, Sir Edward Belcher was sent out to make further researches in Wellington Channel, and in the autumn of the same year the "Prince Albert," touching again at Beechey Island, brought home the joyful intelligence that he had already passed up the channel in open water; and a month later Captain Inglefield arrived in the "Isabel," with the additional and satisfactory information that Sir Edward Belcher had not returned, and had, therefore, probably met with no obstacles.[28] Both ships brought home despatches and letters, showing that additional supplies were urgently wanted, if the objects of Sir E. Belcher, and those of Captain Kellett, to the west, were not to be brought to a premature conclusion;[29] and, accordingly, last summer large reinforcements were sent out in the "Phoenix" steamer, under Captain Inglefield, for the use of both branches of the squadron.

When this officer left the depôt at Beechey Island, on 24th August last, it was known that Sir Edward Belcher had verified the existence of the open sea to the north and north-west beyond Wellington Channel; and it was also known that Captain Kellett had dispatched a foot and sledge expedition, prepared for an absence of 90 days, across Melville Island to the north and north-west, with the / view of exploring such portion of the northern shores of this land as could be effected with the resources at command.

This exploration, under the leadership of that most able and energetic officer Captain M'Clintock, is one of the utmost importance, as being in the right direction; but, when Captain Inglefield sailed from Beechey Island, its result was not known, the period not having elapsed for Captain M'Clintock's return.

This is the latest intelligence we have from the Arctic regions, and this, alas! is the moment chosen by your Lordships for pronouncing authoritatively on the fate of the absent.

It is true that Captain Inglefield brought home also, in the "Phœnix," the news from another branch of the searching squadron, of the discovery of the North-west Passage. It was great and welcome tidings, of itself, perhaps, a sufficient compensation for all the pains and all the expenditure bestowed, with exclusive intention, as we have your Lordships' authority for stating, on the other and still nobler cause of humanity. And yet the solution of the geographical problem appears to have sealed the doom of my unfortunate husband and his brave associates; of those without whose self-sacrifice for the same object, in the fulfillment of their duty, this geographical problem might never have been solved at all.

The intelligence of the North-west Passage was brought home in October last, and before the close of the year the removal of the names of the officers and crews of the "Erebus" and "Terror" was, if I am not misinformed, under deliberation, and was confidently announced in the "Times," which paper, notwithstanding that your Lordships assured me at the time that it had no authority in the statements, proved to be the correct exponent of your Lordships' sentiments.[30]

My Lords, I cannot but feel that there will be a stain on the page of the Naval Annals of England when these two events, the discovery of the North-west Passage, and the abandonment of Franklin and his companions, are recorded in indissoluble association.

It is with reluctance I have spoken of my own efforts, for the purpose of proving that from the beginning, and not recently only, and in consequence of the failure of expeditions in other parts, or from an insatiable desire for random explorations, have I urged upon your Lordships the examination of the Northern Sea, its coasts and islands, between Wellington Channel and Behring Strait, or beyond, as the quarter where the missing ships and crews have yet to be looked for.

The expedition of the "Isabel" screw steamer, which, for two successive seasons, it has been my endeavour to send to Behring Strait, has had the same object in view, namely, that of entering, though at the opposite extremity, this unexplored region, or at least of discovering a channel into it. Yet your Lordships are well aware that it was only as, year after year, my entreaties that you would yourselves send effective steam-vessels to this quarter were unavailing, that I felt myself forced to resort to my own feeble resources.

The expedition of the "Isabel" met with the most cordial support and approval from the President and many distinguished members of the Geographical Society,[31] from the Hydrographer of the Admiralty,[32] and from many Arctic officers, including all those whose experience in Behring Strait gave them the best title to judge its utility. I was assured by the latter, that if I could succeed in getting my little vessel to reach the field of search in season, her services might be valuable.

Nor did your Lordships, though declining to take the measures which were recommended to you at this period, refuse to help me in my own.[33] Without your kind aid, or that of your predecessors in office in 1852, I could not have obtained possession of the vessel without the facilities which last year you kindly gave me for her outfit; and especially without that valuable document addressed to your officers in the Pacific, which seemed to promise all the aid that could be required in time of need, I could scarcely have ventured to send the "Isabel" to sea.[34] /

My Lords, I felt grateful for these benefits; yet, if I could have foreseen that on the first emergency, and when the greatest difficulties of the outward voyage were already passed, you would have denied the interpretation which, in the opinion of the Commander-in-Chief in the Pacific,[35] it was incapable of, and which would have enabled the "Isabel" to carry on her mission, it would have been a kinder thing to refuse me all. Much toil, anxiety, and money would then have been spared, and the vessel would not now be lying at Valparaiso, a monument of my own blighted efforts and of your unlooked-for desertion.[36]

I have purposely omitted, as in no way affecting the question of the necessity of such an expedition as that in which the "Isabel" has failed, the researches of the "Enterprize" and "Investigator," in their course eastward from Behring Strait to Melville Island.[37] In this course, Captain M'Clure has been fortunate enough to find that much-desired link between previous discoveries on the east and on the west, which, like the keystone of the arch, binds the labours of former and present generations together. But as affects the fate of the missing expedition, his brilliant discovery leaves it untouched. Indeed it might have been confidently predicated of those ships, by those who knew the results of Captain Austin's expedition, which were unknown to their commanders, that on the route the "Investigator" was taking, and in which the "Enterprize" appears to be following her, the missing navigators could never be found. Hundreds of miles and intervening land separate the courses north and south of the opposite navigation.

III. Notwithstanding, the experience of Captain M'Clure, during his voyage and long detention in the ice, is not devoid of important deduction. It proves that men may be absent in Arctic climes three or four years, and need not on that account be given up for lost; nay, that they may return in full health and vigour, thus adding new force to the remarkable fact, that the loss of life in the Arctic service, whether arising from casualty or disease, is less than in any other part of the globe where our Navy is employed, in spite of all the hardships and dangers which necessarily attend it. To bring forward the unknown fate of the "Erebus" and "Terror" in opposition to this statement, would evidently be a begging of the question.

Rather may it be asserted, that justice has not been done to the favourable side of the argument. We know that Captain M'Clure was safe and well after nearly four years' endurance of icy imprisonment;[38] that Sir John Ross, under much more unfavourable circumstances, and after a somewhat longer absence, returned home in safety to tell his fate;[39] and that the four Russian sailors, thrown without resources of any kind upon the coast of Spitzbergen, were not liberated till after the lapse of between six and seven years.[40] But shall it be affirmed in any of these cases, and especially in that of the Russian sailors, who endured the longest, and were yet found in health and good condition, that the moment of their liberation or rescue was the utmost term to which their existence would have been prolonged? And yet, such is the conclusion too often involved in the reasonings upon this subject, and because at the end of seven or eight years the crews of the "Erebus" and "Terror" have not been rescued or liberated, or heard of, they are considered to be dead! Indeed I think this a rash and unjustifiable conclusion, and that it would be more reasonable to argue, that if they could live six or seven years, as the Russian sailors did, who were still in health and vigour, they might live double that period, or more.

It will be said we do not know that they have lived seven years. No, my Lords, but we know nothing to the contrary, and it may be that the reason we do not know they are living men, is because your messengers of mercy and deliverance have not been to the spot where alone the truth is to be found, for their voices cannot reach you across the waste of waters, and they are helpless to extricate themselves. Captain M'Clure was shut up in a position from which he could, by abandoning his ship, fall back upon an inhabited coast, or advance on foot to a depôt of provisions of which he knew the existence at Port Leopold. Sir John Ross also was not so far distant from the fishing grounds of the whalers but that he could risk embarking in an open boat to reach them. The Russian sailors were on an island visited annually by ships; but the lost crews of the "Erebus" and "Terror" are presumed of be in a part of the Arctic seas where, having lost all locomotive power, they may hoist their signals of distress in vain. / And if this be their position, it may well move our deepest compassion, but it is not such as should lead us to despair of their prolonged existence.

One of the most experienced of Arctic explorers has consolingly assured us, that life may be maintained in the farthest Arctic lands under circumstances at first sight seemingly the most hopeless; and I believe the same accurate and philosophic observer has remarked, how readily nature accommodates herself to circumstances, and that the hardships and sufferings of the first years would be mitigated afterwards.[41]

But still farther. There are grounds for hope that in the high latitudes, where we believe our exile-navigators to be imprisoned, a dreary existence may be rendered more supportable by a climate of less intense severity, and by an increased abundance of natural resources. Even in the lower latitudes with which we are best acquainted, both sea and land are described, here and there, as "teeming with animal life;" but as Dr. Kane remarks, "at the utmost limits of northern travel attained by man, hordes of animals of various kinds (including the ruminating animals, whose food is a vegetation), have been observed travelling still further north. Birds, of which such almost incredible numbers are occasionally seen, take their flight northward, and the highest waters yet attained are frequented by the whale, the walrus, and the seal, which furnish not only food, but fuel and clothing."[42]

The experience of Captain M'Clure adds something to the mass of facts we already possessed on this subject, for he found a large island where only a blank space existed on our charts before, abounding in game of the noblest description, and supporting a race of well-conditioned and contented Esquimaux, whose existence in that part had never been suspected.[43] Why should not such another island, or more than one, be found in that northern space, the white paper of the charts, of the nature of which we at home know nothing, though it may even now be the abode, and if not timely rescued by Divine or human interposition, may become the grave of our hapless countrymen?

Again, the experience of Captain M'Clure seems to add something also to our evidence in favour of a less inhospitable temperature in the north, for he tells us, though apparently without drawing any conclusion from the fact, that whenever the north wind blew it was warmer.

Nor should it be forgotten, in enumerating the elements of a reasonable confidence in the prolonged existence of the absent voyagers, that they were most abundantly supplied with ammunition, and that, as Captain Penny has judiciously observed, they were all the more likely to be preserved in health, because they would have to seek their subsistence, and thus have their minds and bodies actively employed.

IV. I shall not trouble your Lordships by dwelling further on this head. My husband's conviction, that where Esquimaux can live, there also can Englishmen, with their superior intelligence and larger appliances, has been often quoted. But it is not so constantly remembered that Englishmen are also more provident than Esquimaux, and that at the very outset of his voyage, and while Captain Fitzjames was writing home, that so deep and heavily laden were the ships, that if they reached the Pacific that year, some of the provisions must be thrown overboard for safety. While this sanguine officer was thus writing, my husband, than whom no one could know better what the day of need might require, was diligently adding to his already abundant stock, by means of the guns of his shooting parties, and contemplating a detention of several years.

The evidence of this latter fact, and of the early means he was taking to provide for it, is undeniable.

The deposition of Captain Martin,[44] of the "Enterprize" whaling ship, who was for some time in company with the "Erebus" and "Terror" in the middle ice of Baffin's Bay, was to this effect: that "Sir John assured him, in answer to questions put to him by Captain Martin, that 'he had provisions for five years, and if it was necessary, he could make spin out seven;' moreover, that 'he would lose no opportunity of adding to his stock.' Other officers made the same declaration, stating also, that the ships would winter where they could find / a convenient place, and in spring push out as far as possible, and so on, *year after year*."

This "solemn declaration" of Captain Martin was made before the provost of Peterhead, who has assured me, as have also several other respectable inhabitants of that place, that he is a man of the strictest integrity, truthfulness, and accuracy.

The declaration of Captain Walker, of the whale ship "Union," who, at the time alluded to, was first mate of the "Enterprize," was also made before the provost of Peterhead, and was precisely to the same effect.[45]

Your Lordships are aware that there are letters from the expedition, dated in Baffin's Bay, of like purport; that from Mr. Blenby, icemaster of the "Terror," to his wife, begging her to let no one dishearten her as to the length of their absence, which might be six or seven years, has been published.[46] Mr. Blenby had shared the rude captivity of Sir John Ross in Regent Inlet, and knew how the long absent may be given up for lost, and yet return again to their homes and country.

These last words of men so full of faith and hope, at a moment when they were about to quit the precincts of the known world and plunge into the unknown, seem to me a touching appeal to the long-enduring sympathy and untiring patience of their countrymen.

And even if their hopes may be considered too high, or that they can be convicted of rashness in entering into those unknown seas (were not all the Arctic seas once unknown?) without any harbours of refuge in advance, or any line of depôts in their rear, without assurance of reinforcements or relief from home, or any promise but that which their own heart-trust in their country, and in you, gave them of being looked after; even if this were rashness, is that a reason to abandon them? They went forth, my Lords, at your bidding, and went to those seas which you gave them liberty to explore; you gave them no restrictions such as have abounded in the orders of those who have gone in search of them; they were not told to spare themselves, not enjoined to run no risks, nor restricted in time, though their mission was evidently thought to be a much shorter and much easier matter than it has proved to be.[47] They were themselves prepared, however, to do a work of unknown difficulty and danger, and I well know were not prepared to return till they had spent themselves in its attainment. They have deserved, surely, I may say, they have deserved of their country that she should ascertain their fate.

And I need not tell your Lordships that to follow them whither they have gone is not to encounter the same dangers that they have done; I could not urge it if it were so. But with such vastly superior ships as you have now in the Arctic seas, provided with powerful steam machinery, and other appliances, with the experience in sledge travelling which has been of late years acquired, and with those large precautionary measures as to depôts in advance and in the rear, which you know so well how to devise, it could, I believe, be done with comparative safety. And doubtless it will one day be done. The most northern portion of our globe will not always be a *terra incognita*. When Arctic expeditions for the sake of the missing navigators have long ceased to be familiar to the public ear, and wars and rumours of wars have passed away,[48] the interest in those geographical and other problems which were left unsolved in the year 1854 will again appear worthy of a great national effort for their solution; and then will arise in touching association, the memory of the men who, in pursuit of this knowledge, and in obedience to their country's command, first penetrated into the fastnesses of the north and were left there to their fate. Perhaps it will be the wonder of that future generation that this should have been done, or that any discoveries of great scientific interest and importance should have been abandoned by the Government at the conspicuous moment when it had at its disposal a fleet of invulnerable ships, fit, and fit alone, for Arctic service, and still afloat in Arctic seas, and a host of trained and brave explorers, better disciplined for their work than ever, a combination such as was never seen before, and may never be seen again.

Pardon me, my Lords, that I express myself thus strongly. I would not appear ungrateful for what has been already done. When I look at that fleet of invulnerable / ships, at that phalanx of gallant and devoted men in hard conflict with nature, yearning for the distinction of saving their fellow men, and consider the generous expenditure and the boundless sympathy which have produced the noble spectacle, I pause, and for a moment doubt whether I should have written as I have done.

And yet it is still true that your noble work is incomplete, and that the glory which has hitherto invested it, is about to set in clouds and darkness. It will remain an imperishable fact that the search for these brave martyrs to their duty was given up, not because every part of the Arctic seas had been searched for them in vain, as is too often asserted, but because you have not distinctly authorised, nor sufficiently enabled them to be followed where alone they are to be sought, with any probability of success. Any attempt to divert men's minds from this melancholy truth will I am sure eventually fail.

It is to record my own dissent from such a fatal conclusion, and respectfully to protest against the arbitrary decree you have announced, that I have thus ventured to address you. Would that others who might prevail with you better than I can do, had rendered my hard task unnecessary; that they could induce you to feel that the blessing of them who were ready to perish might yet be yours.

My advocacy must be weakened, perhaps even my facts suspected, because I am too deeply interested, and indeed in some respects my position is a false as well as painful one, for as I could not have dared to plead with you at all unless I had had a husband's life at stake for my excuse, so it may look as if for his sake alone I pleaded, and expected such great things to be done.

There are some I trust amongst those who share with me a common sorrow who will not judge me thus, and all I think must feel, that had my humble endeavour met with any measure of success, it would have been for the good of the whole, as well as of him whose name has sometimes been too exclusively used as the representative of a corporate misfortune.[49]

As to the approbation or the censure to which any poor efforts on my part have been obnoxious, my heart has been too full, and is so still, to be either oppressed by the obloquy or elated by the praise.

It remains for me only to thank your Lordships for the communication you have been pleased to make me, that the widows of those who are to be considered as having died in the service of their country, after the 31st of March next, will be entitled to pensions, according to the existing regulations. Your Lordships will scarcely require me to tell you, after what I have written, that I do not feel it in my power either to claim or to accept a widow's pension.

Before concluding this long and painful letter, allow me to express a hope that I have not now, nor at any time, abused the privilege which belongs to weakness and irresponsibility, or which has been accorded to me by your generous indulgence; and if any hasty expression such as I ought to have avoided has escaped my pen, I entreat you to overlook it, as not intentionally disrespectful.[50]

I have, &c.
(signed) *Jane Franklin.*

17

To: BENJAMIN DISRAELI
4 Spring Gardens, London, [Sunday] 2 April [1854]
Source: Original, H, A/IV/N/14
Editorial comment: Dating: Year was determined by comparing the information in this letter to others in the collection. The year has been added, in a different hand, to the top right-hand corner of the letter and is proved correct by context. [*sic*]: independant.

4. Spring Gardens
2d April

Dear Mr Disraeli

May I venture to ask of you the great favor of bestowing a glance, however hasty, at a letter of mine[1] which has just been printed on the motion of Sir Thomas Acland?[2]

I think I have a strong case against Sir James Graham in his premature removal of the names of my husband & his companions from the Books of the Admy, whilst the ships sent out by his predecessors are still engaged in the search.[3] But since this arbitrary measure cannot touch the truth, it is only because we must regard it as a prelude to the abandonment of all further effort & the consequent immediate recall of the ships, that it affects our interests so fatally.

Sir Th. Acland has given notice of a motion on Tuesday next, founded on my letter, and I venture to hope you may find it possible to support him by a few words – it would confer on me a very great obligation – Our object is to / avert the sending out preemptory orders for the recall of the ships until some thing has been done for the exploration of a quarter which has not yet been examined. – The Admiralty have not attempted to gainsay the facts and conclusions in my letter which has received no other answer than a formal acknowledgment.

I need not assure you that independant of the personal favor you would do me by supporting Sir Th. Acland who is no speaker, it would please him much to be assured of your help.

It has been told me that Lord Stanley[4] takes considerable interest in Arctic matters. I do not venture to write to him, but should you be aware of any kindly disposition on his part, perhaps you would not object to take advantage of it in my behalf at so critical a moment.[5]

The Admiralty are conscious of two weak <or assailable> points in their late proceedings
– the one is the not waiting for the result of Captn McClintock's exploration
– , the other, is the cruel breaking of faith with me in the affair of the "Isabel." They are both alluded to in pages 8 & 9 of the letter of which I take the liberty of sending you a corrected copy, in case your own should have been thrown aside.[6]

Pardon my intrusion for the sake of all that / I have at stake & believe me

dear Mr Disraeli

very truly yours
Jane Franklin

The Right Hone B. Disraeli

18

To: BENJAMIN DISRAELI

4 Spring Gardens, London, [Tuesday] 27 April 1854

Source: Original, H, A/IV/N/7

Cover: Lady Franklin. | The Right Honorable | B. Disraeli MP.

Editorial comment: The year appears to have been added later (possibly by an archivist), but is proved correct by context.

4. Spring Gardens
April 27. 1854

Dear Mr Disraeli

Profiting by your kind suggestion that the Arctic discussion might be revived when the Instructions are produced <which we expect will be instantly> I venture to make a comment upon their anticipated contents.[1]

They will I am told give discretionary power to the Commander of the squadron for one ship to remain out for one more winter. – Such a measure would, if literally acted upon, be quite nugatory – 2 ships, it is well known are essential, the one being a steamer, and it is much to be desired that the discretionary power should not be confined to the Commodore Sir E. Belcher who was in ill health when he left England & whose decision to come home <with his ships> we could pronounce before hand, but that he should be required to admit to his counsels the commanders of the 3 other ships, viz. Captains Kellett, McClintock & Osborn, together with Captn Inglefield who is going out in the / Phœnix.[2]

I have seen a note from Sir James Graham to Sir Roderick Murchison which states that he is not prepared to let the search be carried deeper into <u>the ice</u> in a higher latitude. Sir James might have said deeper into the open water for any thing he or any of us know to the contrary, but this prohibition is fatal to any useful result.

However – , tho' we may not get the instructions mended, yet if an adequate expression of opinion <& feeling> be elicited in the house, it will influence the deliberations on the spot & strengthen the hands of the officer left in command. One of the officers I have mentioned has written home to his friends that he would rather be left behind on half pay than be forced home with the rest & I believe that he would find volunteers to remain with him in spite of the attractions of the Black Sea & the Baltic.[3]

I feel sure that you will be kind enough to move Lord Stanley to give us his kindly promised support at the right moment.

Sir Thomas Acland will be in town tomorrow[.]

Believe me yrs very sincerely & obliged
Jane Franklin

The Right Honble
B. Disraeli

19

To: [LORD PALMERSTON]
60 Pall Mall, London, [Tuesday] 26 June 1855
Source: Handwritten copy, H, A/IV/N/9b
Editorial comment: Written in the same hand as H, A/IV/N/5a, b, and c. [*sic*]: there are no possessive apostrophes in this transcript.

Copy

60 Pall Mall
June 26. 1855

My Lord

As the widow of Sir John Franklin I trust I may be excused for intruding a few moments on your time, in reference to the Arctic Select Committee of the House of Commons, which has received your Lordships sanction as First Minister of the Crown.[1]

Sir John Richardson in a letter published in the Times of the 23rd has claimed for the Erebus and Terror under the command of my late husband, the first discovery of a N. W. Passage, and he treats this as a fact beyond dispute.[2]

That this fact has not been earlier & generally recognised is doubtless owing to the discoveries of Captain McClure having become known in this country above a year before the arrival of Dr Rae last autumn. / It was the testimony of Dr Rae confirmed by the indisputable relics he procured & brought home, which proves that a large body of men belonging to the Erebus and Terror were seen alive on the North American coast in the neighbour-hood of the great Fish river in the Spring of 1850, and this fact, in connection with the other circumstances referred to by Sir John Richardson is, though not the accomplishment of the N. W. Passage, nothing less than the discovery of it.

Captain McClures discovery of another passage which he also was unable to accomplish, in a remoter quarter of the Arctic Sea, was made 6 months later than the above. The one left his ship (the Investigator) inextricably fixed in the ice, a perishable trophy of his memorable enterprise, – the bleached bones of the others (the crews of the Erebus & Terror) lie unburied / on the shore, sad but conclusive memorials of a noble discovery, which had already cost those brave men 5 years of unknown trial & suffering.

They have not perished in vain, – but they are beyond the reach of their countrys rewards & honors, and the humble individual who now addresses you, has nothing to gain, and nothing to desire, in advocating the claims of those who cannot speak for themselves, beyond the recognition of a truth honorable to the memory of the dead, and which detracts nothing from the merits of the living.[3]

Captain McClure and those other brave & distinguished men who have risked their lives in the service of humanity, and who in so doing have greatly extended our knowledge of the earths surface, & helped to set at rest the problem of centuries, are well entitled to all the rewards their admiring / country can bestow, and have an especial claim on my own deep and life long gratitude.

I believe I should not have ventured in troubling your lordship with this letter could I have felt sure that Sir John Richardsons letter which had escaped my own observation, and another letter which appeared in the Times of yesterday from Dr Rae, would come under the notice of the Committee appointed to investigate the claims of Captn McClure.

The authority of Sir John Richardson, himself one of the most distinguished of Arctic discoverers, and the first to devote himself to the search of his friend & former companion in peril, is I am well aware, entitled to a consideration which ought to make any observations of my own, superfluous.

I have the honor to be

my lord

Your lordships very obedient servant

(signed) Jane Franklin

20

To: BENJAMIN DISRAELI
60 Pall Mall,[1] London, [Monday] 2 July 1855
Source: Original, H, A/IV/N/9
Editorial comment: [*sic*] Captain Maclure, Maclures. Lady Franklin's consistent misspelling of Captain McClure's name implies her dismissal of him, even while she acknowledges his discovery.

60 Pall Mall
2^d^ July 1855

Dear Mr Disraeli

I have met with so much generous kindness from you on former occasions that I venture once more to solicit your attention to the subject of my present anxiety[.] It is connected with the Select Committee of the House upon the Arctic question & I am most anxious thro' you to engage Lord Stanley's attention to the point.[2] The inclosed letter of Sir John Richardson published in the Times of the 23^d^, that of Dr Rae in the Times of the 25th & my own letter to Lord Palmerston founded upon these / make it unnecessary, <for me> if you are kind enough to glance at them, to enter into any further explanation.[3] I will only add that the NW. passage discovered by my husband & his companions & for which they laid down their lives was not only discovered by them earlier than the present claimant's, but that the passage thus discovered is a navigable one ~~with~~ as the tides & the <pieces of> wreck floated on these tides abundantly prove,[4] whereas the other passage discovered 6 months later by Captn Maclure is we have reason to believe from its narrowness & the accumulation of ice at its N. end, eternally blocked.

That my husband's discovery was not perceived & acknowledged from the moment of Dr Rae's return last autumn is not <perhaps>

much to be wondered <at> as the report of Captn Maclure's discovery brought home by Captn Inglefield had preceded by a whole year the arrival in England of Dr Rae & thus Captn Maclures fame had already been established as the Discoverer of the N. W. passage & justly so, till the revelations of Dr Rae / altered the aspect of affairs.[5] There are few however who understand the subject, sufficiently to discriminate facts, & my own voice was silent under the crushing misery of Dr Rae's tragic story.[6] It was the letter of Sir J. Richardson alone which stimulated me to make an effort ~~an effort~~ on behalf of those whose voices are hushed for ever, but who seem to have an especial claim on our sympathy, as their ~~their~~ triumph if indeed it may be so called, was purchased by such sufferings as have left them lifeless & unburied on the shore. Surely the dead should not be grudged this justice & this honor to their memories & I cannot believe that it will be left to my own feeble pen or to some other as the chronicler of my husband's life to claim <tardily> for him <& them> that <justice> which would be more fitly awarded on higher authority. I am however supported by the highest, that of many other Arctic navigators who coincide with the view now brought forward by Sir J. Richardson, tho' acknowledging that they had not given it attention before. Captn Collinson the Chief of Captn Maclure's expedition & who has himself / claims of the highest order to consideration has been hardly used by the terms of the notice of the Select Committee which not only I believe ignores any other discoverer of a NW passage than Captn Maclures, but seems to ignore also that he <Captn C.> was chief in Command of that expedition.[7] I venture to presume that you will do me the favor of drawing the attention of Ld Stanley as a member of the Committee now sitting, to the subject of this letter, hurriedly written in consequence of hearing from Sir Th. Acland who is absent, that the committee was to meet to day.

Pardon my intrusion & believe me dear Mr Disraeli

yrs ever truly & obliged

Jane Franklin

I had a note from Sir J. Richardson on Saturday which I inclose. If it can be returned to me, I shall be glad, as it is an important testimony.[8]

21

To: WILLIAM A. MACKINNON[1]
60 Pall Mall, London, [Friday] 6 July 1855
Source: Printed transcript, *HCSP*, 1854–55/VII/18, 55 (printed twice)
Heading: LETTER from Lady *Franklin* to *W. A. Mackinnon*, Esq., M.P., Chairman.
Editorial comment: Lady Franklin's letter was published twice in the Arctic Select Committee proceedings to adjudicate Captain McClure's reward for his discovery of the Northwest Passage: in the introduction and in the appendix.

Sir, 60, Pall Mall, 6 July 1855.

I VENTURE to trespass a few minutes on your time and that of the Committee over which you preside, in behalf of the claims of my late husband, Sir John Franklin, and his companions, as connected with the subject you have under discussion.

When it is remembered that these brave and unfortunate men, after years of intense privations and suffering, were found dead of starvation upon a spot which they could not have reached without having first solved that geographical problem which was the object and aim of all these painful efforts, and when it is also remembered that they are beyond the reach of their country's rewards, you will not, I think, refuse them the just acknowledgment that is due to their memories.

It would ill become me, and is far indeed from my wish, to attempt to question the claims of Captain M'Clure to every honour his country may think proper to reward him. That enterprising officer is not the less the discoverer of a North-west Passage, or, in other words, of one of the links which was wanted to connect the main channels of navigation already ascertained by previous explorers, because the 'Erebus' and 'Terror' under my husband had previously, though unknown to Captain M'Clure, discovered another and a more navigable passage; that passage, in fact, which if ever ships attempt to push their way from one ocean to the other, will assuredly be the one adopted. And it can never be denied to Captain M'Clure that he is the first who has by his own skill, and by means of the timely assistance of the brave men who were in search of him, made his individual way from one ocean to the other. Such a transit, though not the object which has engaged the attention of the civilised world for centuries, is a distinction of which any man may well be proud.[2]

What I presume to claim for those who can urge nothing for themselves is the first discovery of a navigable passage for ships in that unknown space which lay between the discoveries of former navigators, for to such connecting channel has the solution of the geographical problem for many years past been reduced. My husband was specially warned by his instructions not to seek it in the quarter where the 'Investigator' lies, lest impenetrable ice should, as was anticipated, arrest his progress, and he found the passage by acting (in conformity with his instructions) on those theoretical convictions which, as Sir John Richardson has shown, he deliberately formed.

Convinced, Sir, that it must be your desire, and that of the other honourable Members of the Committee, to do justice to the dead, while you duly and generously honour the living, and believing that these two objects do not clash, but may be harmoniously combined, I have presumed thus to address you.

I trust you will pardon the Widow and the Friend this last effort in behalf of those who have nobly perished.

I have, &c.,

W. A. Mackinnon, Esq., M.P. (signed) *Jane Franklin*.

22

To: LORDS OF THE ADMIRALTY
60, Pall Mall, London, [Saturday] 12 April 1856
Source: Printed transcript, NLC, Mic.F. CC-4 no. 17703 (a copy of *HLSP*, 1856/X/366–71)
Heading: (in HLSP copy only) No. 49. | *Letter from Lady Franklin. – (Received April 16.)* | LADY FRANKLIN presents her compliments to the Secretary of the Admiralty, and requests he will do her the favour to place the accompanying letter before the Lords Commissioners of the Admiralty at the earliest opportunity. | 60, *Pall Mall*, | *April* 18, 1856. | Inclosure 1 in No. 49. | *Lady Franklin respecting the adjudication of the Reward, to the Lords Commissioners of the Admiralty.*
Editorial comment: The private printing (NLC) includes a copy of the *Gazette* notice of 22 January 1856. Written on top of front page in an unknown hand (most likely an archivist's) (NLC copy only): "12$^{\underline{50}}$ | Privately Printed | 551 | NW 998 F8322." Handwritten on the back cover (writer also unknown): "Lady Franklin to the | Admiralty | April 12/56"; plus a stamp of a presumed past owner: "DR. CRANDON, | 366, COMMONWEALTH AVE | BOSTON | JUL 24 1920."
[*sic*]: opposes, woful, forsee, preesnt, Maclure. These appear in both copies of the letter and may all be typesetting errors.

My Lords, 60, *Pall Mall, April* 12, 1856.[1]

I HAVE the honour of laying before your Lordships a copy of the notice published by your orders in the *Gazette* of the 22nd of January last, and of soliciting your consideration of the motives which compel me respectfully to enter a protest against the early adjudication therein proposed.[2] I should not have delayed until this late moment the execution of an intention formed soon after the publication of the notice and communicated many weeks since to

Dr. Rae, but that I have been incapacitated from writing by serious illness, from which I am indeed but partially recovered.[3] /

The grounds upon which I venture to appeal to you against the present adjudication of the reward are the following: –

1. Because it cannot truly be affirmed of the crews of the Erebus and Terror that their fate has been ascertained, and therefore to adjudicate on such an assumption seems to be premature.

2. Because Dr. Rae has not by his efforts ascertained their fate.[4]

3. Because it may yet happen that their fate will be positively and fully ascertained, and if so, those who by their efforts succeed in ascertaining it, will, should a premature adjudication take place, be deprived of their legitimate reward.

4. Because the adjudication, by affirming that there is nothing more to be disclosed, opposes a check to any further efforts for ascertaining the fate of the expedition, and appears to counteract the humane intention of the House of Commons in voting a large sum of money for that purpose.[5]

I trust that in making some observations in explanation of these points, I shall not fail in the respect due to your Lordships, and that you will accord me that indulgence which the subject and my own deep personal interest in it may appear to claim.

1. It is not my intention, under the first of these heads to throw doubt upon the reports brought home by Dr. Rae, confirmed as they are, in some points, by the indisputable relics procured from the Esquimaux, but only to point out that they are convincing within certain limits only; and that much that has hitherto been considered as established, is only conjectural, and should lead to a suspension of judgment, rather than a hasty enunciation of it.[6]

It is not proved, by any facts we are in possession of, that the party of white men who arrived with their large boat (the remains of which, with many articles belonging to it have been found) within the estuary of the Great Fish River,[7] and who are said to have perished there, were the only survivors of the crews of the Erebus and Terror, and that no other remnant of the original ships' companies, amounting to about 135 men, took a different route.

And even as to this (known) boat party, it has not been traced back to the ships or to the wrecks from which it was equipped; nor have the ships been sought for at all, though there is much reason to conclude from the nature of the objects brought home by Dr. Rae, and of others seen by Mr. Anderson on Montreal Island and the adjacent shore, that they had been pillaged by the Esquimaux, and were not far distant.[8] What secrets may be buried in the graves of our unhappy countrymen or in caches not yet discovered we have yet to learn. The bodies and the graves which we were told of have not been found; the books (journals) said to be in the hands of the Esquimaux have not been recovered, and thus left in ignorance and darkness with so little obtained and so much yet to learn, can it be said and is it fitting to pronounce that the fate of the expedition is ascertained?

That your Lordships did not consider that this question was resolved by Dr. Rae's reports at the close of 1854, and by the relics which to a certain extent authenticated them, is shewn by your own proceedings when that tragic intelligence arrived, for it was immediately decided that steps must be taken to verify the truth of these reports which could not be accepted as conclusive, and that further intelligence must be sought for.[9] There was but one feeling in the country on this sad occasion. No amount of expense would have been grudged to make a final expedition of search complete, for it was felt that after six long years of failure and disappointment, the clue which we had asked and prayed for was now in our hands, and that England's honour and credit were concerned in holding it fast and following it up till it led to the solution of the mystery.[10]

My Lords, I shrink from recalling the pain and woful disappointment I felt, and which many others felt with me, when the response to this generous excitement in the public mind, and the sole result of your deliberations, was no more than a birch bark canoe expedition down the Great Fish River, confided to the Hudson's Bay Company, but unsustained by any naval resources. In vain was it pleaded that a vessel might be sent to co-operate with this river party, who, if they ever reached the sea, could not venture to embark upon it in their frail canoes: and, if this were not granted, that at least a naval / officer might accompany and direct the expedition, since it was well known that the Hudson's Bay Company, with all their zeal to accomplish the objects required of them in the most effectual manner, would not be able to supply to it an officer competent to make the indispensable observations for latitude and longitude. To the credit of Dr. Rae and of the Hudson's Bay Company's officers about to be employed, I may observe that he made a similar recommendation, being persuaded that those brave and right-minded servants of the company would not hesitate to place themselves under the leadership of an officer in Her Majesty's Navy, provided he were one already tried and distinguished in Arctic service. To add to the original deficiencies of this over-land or river expedition, it failed to secure an interpreter, so that all the information it has brought back from the Esquimaux, and that derived chiefly from a few women, was transmitted only by signs.[11]

Every praise is due to the exertions of the two zealous officers, who, under all these disadvantages, were able to accomplish anything: but it is scarcely to be wondered at if, after a rapid survey of 9 days only, within a very limited district, which did not extend even to King William Island where our fugitive countrymen were first seen, they were compelled by the state of the damaged boats hastily to return, and have thrown no new light upon the history of those whose fate they went to ascertain. Mr. Anderson has been able to confirm the evidences of a large party from the Erebus and Terror having arrived from the sea within the estuary of the Great Fish River; but his negative testimony on other matters, such as the bodies and the graves which were not to be found, tends rather to throw doubt upon than to confirm them.

I may here, perhaps, be allowed to add, without prejudice to that excellent servant of the Hudson's Bay Company, Mr. Anderson, that *he* is so far from considering, the fate of the expedition has been fully ascertained by the results of his late survey, or that there is nothing more to be done, that he has felt it to be a duty to express to me since his return, as he had done before he started, his decided opinion that a vessel should be sent out to the vicinity of King William's Land to pursue the search. With about twenty men, well armed, and two interpreters, he considers that two parties might be dispatched from it to explore the east and west sides of Victoria Strait, as well as the lower part of Regent Inlet, his own opinion being that the wrecks of the vessels are to be found in

Victoria Strait, on the west coast of Boothia, between Bellot Strait and King William's Land. On the 15th of September last, Mr. A. writes to me, that he had himself made preparations for a second season of search; but, I presume that, having no instructions on this head, they have not been turned to any account. It is evident he considers, that the Esquimaux tribes on the shores of the strait hold the secret we are in search of, and that something more than a flying visit of a few days must be effected, before their, perhaps guilty, fears can be allayed, and their confidence won.

The view I have thus ventured to submit, that there is much more conclusive evidence to be yet obtained as to the fate of the expedition, and that the means of obtaining it had not been used, but are within our reach, is not confined to myself alone, in which case I might, notwithstanding my deep personal interest in the matter, hesitate to advance it. It would be in my power to adduce the most positive evidence that the sentiments and views I have taken the liberty of expressing, are shared by some of the highest authorities to whom your Lordships have been accustomed to refer in Arctic matters, as also by those brave and experienced officers who have most distinguished themselves in Arctic service; and I would entreat of you, before you place an extinguisher upon the light which has arisen in that dark corner of the earth, whither we have been directed as by the finger of God, that you will, as you have done before, call together those Arctic officers, and obtain their individual and collective judgment in this emergency.

2. It seems almost superfluous to observe, if the fate of the crews of the Erebus and Terror has not been ascertained, that Dr. Rae has not succeeded in ascertaining it. And thus I might be spared the ungrateful task of considering whether or not he made the efforts implied in your Lordships' proclamation as a condition of reward, were it not that, by passing over this clause, I seem to admit an assumption that endangers the main truth I wish to establish. It is with great reluctance that I find myself obliged to contest / Dr. Rae's claim to having, by virtue of his efforts, ascertained the fate of my husband and his companions, for we are indebted to him (short of this), for such valuable information, supported by such tangible proofs of a few facts, as might have enabled us, had ampler

means been employed, to set the question at rest, and as enables us still to do so. Had Dr. Rae verified some of the reports he received at second or third hand from the Esquimaux, by personal investigation, and made use of the facts thus attained to get at further truths, or even, had he hastened home with no other object than to provoke the organisation of a much more complete and effective search than any he could himself accomplish on the spot, his claims would bear a different aspect. But he did not go out of his way to test the startling facts communicated to him, and he returned home, as he expressly tells your Lordships in his official letter, to stop further expeditions (in other parts) – a praiseworthy object, perhaps, but one widely different from that for which the reward was offered. It is but due, however, to Dr. Rae, to add that, when the Fish River Expedition was resolved upon, he gave his ready advice for its organisation, though he declined the command of it, which was proposed to him. He also declares himself favourable at this moment, to farther search, were it only, as he has assured me, to secure for his statements that confirmation which he anticipates. Much more gladly would I plead, did I not feel that it is beyond my province, than oppose his right to that which is put forth in the third clause of the *Gazette* notice, to which it appears to me he has not made good his title.[12]

3. For the sake of those who may yet advance a stronger claim to it, I am again compelled respectfully to protest against the premature adjudication contemplated in the *Gazette* notice. When in the early part of the Session of 1849, the House of Commons unanimously voted £20,000 for the encouragement of private enterprise and of private enterprise alone, in the search of the lost objects of national solicitude, they placed no restrictions as to the period during which the reward was to be held out, it could never have been contemplated that while the community was divided in opinion as to the evidence of facts, and while a great majority of those best informed on the subject and others the most deeply interested in it, were dissatisfied with that evidence, and demanded better, an arbitrary edict such as that involved in the premature adjudication of the reward, should, in favour of any one candidate of doubtful pretensions, shut out all future ones. I would fain submit to your

Lordships, that so long as private funds are embarked in the same cause, and that active measures are in progress or in contemplation, to clear up the mystery as to the fate of the crews of the Erebus and Terror, it would be unjust to place an arbitrary limit to the operation of an act designed expressly for that object, and so proclaimed at the time "By Her Majesty's Government."

And here I feel compelled to state that, though it is my humble hope and fervent prayer that the Government of my country will themselves complete the work they have begun, and not leave it to a weak and helpless woman to attempt the doing that imperfectly which they themselves can do so easily and well, yet, if need be, such is my painful resolve, God helping me.

Should these last of the explorers accomplish any of these things, will you ignore their services and claims, because, in the spring of 1856, you had deemed that the fate of the expedition was ascertained, and had given away the reward? I have a right to use this argument, though in carrying it into its consequences I feel it may be doing injustice to your Lordships, as well as to the devoted volunteers who will undertake what they have to do in a higher spirit than any hope of pecuniary recompense can put into them.

But I may illustrate the argument by a more immediate example – that of the zealous and enterprising Captain Penny. It may not be known to your Lordships that Captain Penny, before his departure from England last year, / in command of two whaling ships, informed me that the reports brought home by Dr. Rae reminded him of some vague rumours of a fight between whites and Esquimaux, which had reached him in Northumberland Sound the preceding season, from a great distance, travelling through successive tribes of natives. And Captain Penny added, that in spite of all the difficulties he foresaw in the execution of his project, it was his intention to engage some of the most intelligent and trustworthy natives, domesticated at his whaling station, to trace back these rumours to their source, whether that source were the catastrophe at the Great Fish River, or any other which may have overtaken a separate portion of the crews of the Erebus and Terror, in some other locality. Now, though I have no great hope (considering the distance to be traversed and other obstacles), that much success will attend Captain Penny's laudable efforts, and am sure that no thought of qualifying himself for the Government reward entered into his speculations, yet, would it be fair to adjudicate that reward at this moment, in the face of such a fact?[13]

I have spoken reluctantly of a private expedition, at my own cost, which, in despair of your Lordships completing the work you have begun, and not till then, I may be forced to undertake. And it may not be superfluous, though I trust it is so, that I should state I can have no personal interest in desiring that the adjudication of the reward should be delayed till the result of that expedition, or of any other be known. Even in the event of the reward being adjudged in whole or in part, to those engaged in my private expedition, this could in no degree relieve me from any portion of my own pecuniary obligations to them, or from any expenditure whatever. My funds since the settlement of my late husband's affairs, are equal to the ample equipment of the Isabel schooner, which is now lying in dock, waiting, at a considerable current expense to me, her possible destination; and, unless these my independent funds should become exhausted, which I do not forsee, I shall not even ask your Lordships for the ordinary pension of a rear-admiral's widow, to which I presume I am entitled.[14] My request to your Lordships will be limited to such assistance as is entirely independent of money, and indeed, to such as I have been assured, on the highest authority, will not be denied.[15]

4. And this leads me to add, as my last ground for remonstrating against the immediate decision which Dr. Rae's claims have given you occasion to announce, that its greatest evil, and that of which I should have the most reason to complain, would be the discouragement it gives to many a noble-minded man, who, in volunteering for this service, desires to do so with the confidence that he has your Lordships' opinion, all has been done that can or need be done; that there is nothing more to learn or nothing worth knowing, or nothing commensurate with the cost and risk incurred (little as that cost and risk would be) to obtain it; and it is to brand with infatuation and obstinacy the feelings and convictions that prompt a different course. But if indeed the object of so many

years' labour and anxiety spent, but not wasted, in wrong directions, is to be that which has hitherto been a nation's duty, be now dwindled to a private concern – at least let me entreat that you will not strive to stifle these last private efforts by any act of yours.

Permit me to add one concluding observation: –

It may be surmised that, at the preesnt moment, I have a new incentive to farther search, inasmuch as justice has hitherto been withheld from my husband and his companions as the first discoverers of a north-west passage, and withheld on the ground that future investigation was necessary, in order to determine the extent to which it might be found that they had carried those discoveries to which they had sacrificed their lives.

Such was the barren, though kindly-expressed result of an appeal, which believing my motives must be above suspicion, I took courage to make to a Select Committee of the House of Commons, appointed to inquire into the claims of Sir Robert Maclure to reward, when I found that the right of my husband's expedition – not to *reward*, but – to the distinction of prior discovery was ignored, or not understood, or forgotten.[16] And yet one would have supposed that the full examination of any conflicting claims lay at the very root of a just judgment, and that those who could not plead their own cause because / their voices were hushed in the grave, would have found an advocate in every man in that assembly, as they did find, I gratefully acknowledge, in some.

But this painful remembrance does not enter, or enters but little into the feelings which prompt my efforts for farther search, which I should not desire the less, even did I anticipate that its consequences might be to dissipate the convictions I now have, instead of to confirm them. Nor, perhaps, should I have touched upon this subject at all, except as leading me with less appearance of presumption to express my opinion, that it is due to a set of men who have solved the problem of centuries by the sacrifice of their lives and in the very act of dying, that their remains should be sought for in the place where they perished, and that, as they assuredly devised some means of preserving from destruction the last words they dictated to those they loved and the records of their five long years of suffering and adventure, the recovery of these precious documents should be the aim of persevering exertion and held out as a fitting object for reward.

The best tribute that could be paid to the first and only martyrs to the great Arctic discoveries of the present century, would be a national and final expedition for this holy purpose. The objections against a useless repetition of the attempt will be unanswerable, when once an adequate effort for the attainment of these objects has been made in vain; and then may England feel that she is relieved of her responsibilities, and can close with honour one of the noblest episodes in her naval history.

I have trespassed long on your Lordships' patience, for which I beg to apologise, while renewing my respectful protest against the impending adjudication of the reward, and my earnest request that it may be delayed till such time as the result of the last expedition be known.[17]

I have, &c.

(Signed) JANE FRANKLIN.

23

To: LORDS OF THE ADMIRALTY
60 Pall Mall, London, [Friday] 11 July 1856
Source: Printed transcript *HCSP*, 1857–58/LX/36–7
Heading: CORRESPONDENCE respecting Her Majesty's Ship "Resolute," and the Arctic Expedition. | No. 20 | *Lady Franklin to the Lords Commissioners of the Admiralty.*
Editorial comment: [*sic*]: an, Osborne. These may be typesetting errors.

My Lords, 60, *Pall Mall, July* 11, 1856.

THREE months ago I felt constrained to address your Lordships, requesting that you would be pleased to delay your adjudication of the reward claimed by Dr. Rae for ascertaining the fate of my husband's expedition, until such time as the result of a more complete and final search could be known.[1] I implored your Lordships to adopt such measures as would set this question at rest: at the same time I was compelled to represent that your refusal to do this, would force upon me the painful alternative of taking the burden of an expedition upon myself, at whatever cost, and under great disadvantage.

To this letter I have not been honoured with any reply, but, notwithstanding, it seemed to me and to others, not unreasonable to interpret your silence in a manner not unfavourable to my wishes, inasmuch as your Lordships were well aware that so long as no adverse decision was announced to me, I was precluded from taking any steps for advancing my private expedition, which depended entirely on the non-adoption of the other. Even when I read in the "Gazette," after two months and more had elapsed, that your Lordships, disregarding my request, had given the reward of 10,000*l*. to Dr. Rae, I was still unwilling to regard this action as an absolute rejection of my petition for further service, since in that light, or with such a bias or object in view, it might have been practicable to announce it at a much earlier period, and thus relieve me from suspense and set my hands free for action.[2] But, besides this, I was aware that a memorial to the same effect as my own petition, signed by the most eminent scientific men in London, and embracing the opinions of all the chief Arctic officers, had been presented to Her Majesty's Government (by whom it was kindly received), and I indulged the hope that it could scarcely fail to receive your Lordships' favourable consideration.[3] Thus between doubt and hope, between occasional misgivings and reviving confidence, but withal in constant and harassing anxiety, I have passed three long months (precious months to me, who required them all for my own expedition, if that great burden was at last to fall upon me), till, at last, the time has arrived when the equipment of a private expedition is no longer possible, and a season of probably unexampled openness for ice navigation is passing away.

I feel sure that if your Lordships would only do me the favour of considering for a moment the painful position in which I have thus been and am still placed, without a single word vouchsafed to me, either to confirm my hopes or to extinguish them, deprived of any means but such as I had a reasonable objection to, of securing public feeling in my behalf, whilst the Arctic papers (including my appeal to your Lordships) which were called for in the House of Commons, continued to be withheld – unable thus to make use of the present or calculate on the future, you would feel that a great hardship, nay, that a great injustice (for such I feel it to be), has been inflicted on me.

Yet, great as this trial has been, it receives aggravation from the knowledge that I am not alone affected by it. I abstain from obtruding on you details of private matters, however they might serve to illustrate this aspect of my embarrassing position, but I feel sure that you will deem it worthy of your kind and serious attention, when I inform you that the distinguished individual who has generously offered me his gratuitous services for the command of my private expedition, should I be unhappily reduced to this extremity, has done so at the sacrifice of all his own professional and private interests, in the purest spirit of sympathy with my anxieties, and of devotion to a holy cause.[4] And I might say much more than this if I felt permitted to do so; your Lordships, however, will, I am sure, perceive that I cannot indefinitely prolong the state of uncertainty in which my noble-minded and generous friend is now placed, and that it is my duty either to release him from his promise, as I would gladly do were I sure that my cause were safe in your hands, or enable him at once to commence independent operation.

Regretting deeply that you have, as I learn, come to a decision adverse to the / immediate starting of a vessel by the eastern route, since I fully recognize the possibility of following my husband's track, on that side, down Peel Channel, I yet may be permitted to express the opinion I have long entertained, confirmed as it is by that of your late eminent hydrographer, Sir Francis Beaufort, and by that of Captains Collinson and Maguire,[5] that the route by Behring Strait, though longer in distance, is of surer and safer accomplishment, and that a vessel dispatched this autumn to Behring Strait would, probably, arrive at the spot to be searched in a shorter time than by the other. Captain Collinson, whose experience is the highest that can be adduced on this point, has no doubt that he could carry even such a heavy-sailing ship as the "Enterprise," without the aid of steam, in one season only, to the very locality where the remains of the "Erebus" and "Terror" are now lying;[6] and while it is, at least, certain that the Esquimaux hold the secret of their fate, and of the pillage they have acquired from the catastrophe.

This opinion of Captain Collinson, as to the facility of a vessel reaching the place of its destination, in one season, by way of Behring Straits, is shared by Captain Maguire, as expressed in a letter which I have permission to inclose;[7] your Lordships will also perceive therein another reason for the adoption of this route, which has not hitherto received the attention its extreme importance deserves, namely, the facility it gives of bringing the vessel into close contact with the Esquimaux, it being Captain Maguire's opinion (as it is that of Mr. Anderson, the late commander of the boat party down the Great Fish River), that the tranquil presence of a vessel is necessary to extract the whole truth from the natives.[8] These people are not wanting in sagacity, and if they see nothing but a boat or sledge party, they will be sure to calculate on the very limited resources of such a party, that it will soon return whence it came, and rid them of unwelcome investigations. It is also to be recollected that the Esquimaux are in the habit of making spring and autumn migrations, so that time would be required to enable the intelligence that white men were on the spot to permeate throughout the country, and thus reach the ears of any stragglers that may yet remain of the crews of the missing vessels.

I would entreat of your Lordships, should you doubt the accuracy of any of my statements, to call before you those two able and experienced officers, Captains Collinson and Maguire, one of whom has brought back his ship and crew in perfect safety, after a navigation in Arctic waters of unexampled length and importance, whilst the other, within a more restricted field of action at Point Barrow, succeeded so well in his endeavours to gain the confidence and co-operation of the natives, as to be an earnest of his success in any other quarter.

I mention these two distinguished officers as being especially qualified to speak of the advantage and disadvantage of the route suggested, not forgetting that Captains Osborne and Richards are also on the spot, equally able to submit to your Lordships, if honoured by your reference, all that might have been said in favour, or otherwise, of the route which you have pronounced to be impracticable at this advanced season.[9] All are alike ignorant that I am expressing this unbounded confidence in their capacity and zeal, in the humble hope of reminding your Lordships that if you give little weight to anything I can advance as coming from an incompetent or too-interested person, there are those at hand whose qualifications, whose duty towards you, and whose sense of responsibility, remove them widely from such disparaging circumstances.

Whilst this subject is still under deliberation, I commit the prayer of my present appeal to your serious and humane consideration, believing that the honour of my country is no less concerned in the result, than in my own personal interests and those of my fellow sufferers in calamity.

I have, &c.

(Signed) JANE FRANKLIN

24

To: LORD PALMERSTON
60 Pall Mall, London, [Tuesday] 2 December 1856
Source: Printed transcript, in F. Leopold McClintock, *The Voyage of the 'Fox': A Narrative of the Discovery of the Fate of Sir John Franklin and His Companions* (London: John Murray, 1859), 352–60.
Heading: APPENDIX. | No. I. | A LETTER TO VISCOUNT PALMERSTON, K.G., &c., FROM LADY FRANKLIN,

60, Pall Mall, December 2, 1856.

MY LORD, –

I trust I may be permitted, as the widow of Sir John Franklin, to draw the attention of Her Majesty's Government to the unsettled state of a question which a few months ago was under their consideration, and to express a well-grounded hope that a final effort may be made to ascertain the fate and recover the remains of my husband's expedition.

Your Lordship will allow me to remind you that a Memorial with this object in view (of which I enclose a printed copy) was early in June last presented to, and kindly received by you.[1] It had been signed within forty-eight hours by all the leading men of science then in London who had an opportunity of seeing it, and might have received an indefinite augmentation of worthy names had not the urgency of the question forbidden delay. To the above names were appended those of the Arctic officers who had been personally engaged in the search, and who, though absent, were known to be favorable to another effort for its completion. And though that united application obtained no immediate result, it was felt, and by no one more strongly than myself, that it never would be utterly wasted.

I venture also to allude to a letter of my own addressed to the Lords Commissioners of the Admiralty in April last, and a copy of which accompanied, I believe, the Memorial to your Lordship, wherein I earnestly deprecated any premature adjudication of the reward claimed by Dr. Rae, on the ground that the fate of my husband's expedition was not yet ascertained, and that it was due both to the living and the dead to complete a search which had been hitherto pursued under the greatest disadvantage, for want of the clue which was now for the first time in our hands.[2]

The Memorial above alluded to, and my own letter of earlier date, had not yet received any reply, when, in the month of July, the Lords of the Admiralty caused prompt enquiries to be made as to the possibility of equipping a ship at that advanced season, in time for effective operations in the field of search. The result was that it was pronounced to be too late, and the subject was dismissed for that season.[3]

Upon this I addressed a letter to the Board (of which I take the liberty to enclose a copy), respectfully showing that by this unfortunate delay the opportunity had also been taken from me of sending out a vessel at my own cost, a measure which I had previously felt myself obliged to state to their Lordships would be the alternative of any adverse decision on their part.[4] I pleaded therefore, as the only remedy for the loss of an entire summer season, that the route by Behring Straits was by some of the most competent Arctic officers considered preferable to the eastern route, and that the equipment of a vessel for this direction need not take place before the close of the year.

In reply, their Lordships caused me to be informed that "they had come to the decision not to send any expedition to the Arctic regions in the present year."

This communication, however, was in answer merely to my own letter. The memorialists had as yet received no reply, and accordingly the President of the Royal Society[5] put a question respecting the Memorial in the House of Lords at the close of the session, which drew from one of Her Majesty's Ministers (Lord Stanley), after some preliminary observations, the assurance that Her Majesty's Government would give the subject their serious consideration during the recess. I may be permitted to add, that, in the conversation which followed, Lord Stanley expressed himself as very favourably disposed towards a proposition made to him by Lord Wrottesley, that, in the event of there being no Government expedition, I should be assisted in fitting out my own expedition; an assurance which Lord Wrottesley had the kindness to communicate to me by letter.[6]

But, my Lord, as nothing has occurred within the last few months to weaken the reasons which induced the Admiralty, early in July last, to contemplate another final effort, and as they put it aside at that time on the sole ground that it was too late to equip a vessel for that season, I trust it will be felt that I am not endeavouring to re-open a closed question, but merely to obtain the settlement of one which has not ceased to be, and is even now, under favorable consideration. The time has arrived, however, when I trust I may be pardoned for pressing your Lordship, with whom I believe the question rests, for a decision, since by further delay even my own efforts may be paralyzed.

I have cherished the hope, in common with others, that we are not waiting in vain. Should, however, that decision unfortunately throw upon me the responsibility and the cost of sending out a vessel myself, I beg to assure your Lordship that I shall not shrink, either from that weighty responsibility, or from the sacrifice of my entire available fortune for the purpose, supported as I am in my convictions by such high authorities as those whose opinions are on record in your Lordship's hands, and by the hearty sympathy of many more.

But before I take upon myself so heavy an obligation, it is my bounden duty to entreat Her Majesty's Government not to disregard the arguments which have led so many competent and honorable men to feel that our country's honor is not satisfied, whilst a mystery which has excited the sympathy of the civilized world, remains uncleared. Nor less would I entreat you to consider what must be the unsatisfactory consequences, if any endeavors should be made to quench all further efforts for this object.

It cannot be that this long-vexed question would thereby be set at rest, for it would still be true that in a certain circumscribed area within the Arctic circle, approachable alike from the east, and from the west, and sure to be attained by a combination of both movements, lies the solution of our unhappy countrymen's fate. While such is the case, the question will never die. I believe that again and again would efforts be made to reach that spot, and that the Government could not look on as unconcerned spectators, nor be relieved in public opinion of the responsibility they had prematurely cast off.

But I refrain from pursuing this argument, though, if any illustration were wanting of its truth, I think it might be found in the events that are passing before our eyes.

It is now about two years ago that one of Her Majesty's Arctic ships was abandoned in the ice.[7] In due time this ship floated away, was picked up by an American whaler, carried into an American port, and (all property in her having been relinquished by the Admiralty) was purchased of her rescuers by the American Government, by whom she has been lavishly re-equipped, and is now on her passage to England, a free gift to the Queen. The 'Resolute' is about to be delivered up in Portsmouth harbor, not merely in evidence of the cordial relation existing between the two countries, but as a lively token of the deep interest and sympathy of the Americans in that great cause of humanity in which they have so nobly borne their part. The resolution of Congress expressly states this motive, and indeed there could be no other, as it is well known that for any purpose but the Arctic service those expensive equipments would be perfectly useless and require removal.[8]

My Lord, you will not let this rescued and restored ship, emblematic of so many enlightened and generous sentiments, fail, even partially, in her significant mission. I venture to hope that she will be accepted in the spirit in which she is sent. I humbly trust that the American people, and especially that philanthropic citizen who has spent so largely of his private fortune in the search for the lost ships,[9] and to whom was committed by his Government the entire charge of the equipment of the 'Resolute,' will be rewarded for this signal act of sympathy, by seeing her restored to her original vocation, so that she may bring back from the Arctic seas, if not some living remnant of our long-lost countrymen, yet at least the *proofs* that they have nobly perished.

I need not add that we have as yet no proofs, whatever may be our melancholy forebodings. That such is the fact, in a legal point of view, is shown by a case now or lately pending in the Scotch courts, in which the right of succession of a considerable property is not admitted, on account of the absence of all but conjectural testimony.[10] In this aspect of the question I have no personal interest, but it is one that may not be deemed unworthy of your Lordship's attention, combined as it must be with the fact that our most experienced Arctic officers are willing to stake their reputation upon the feasibility of reaching the spot where so many secrets lie buried, if only they are supplied with the adequate means.

It would be a waste of words to attempt to refute again the main objections that have been urged against a renewed search, as involving extraordinary danger and risking life. The safe return of our officers and men cannot be denied, neither will it be disputed that each succeeding year diminishes the risk of casualty; and indeed, I feel it would be especially superfluous and unseasonable to argue against this particular objection, or against the financial one which generally accompanies it, at a moment when new expeditions for the glorious interests of science, and which every true lover of science and of his country must rejoice in, are contemplated for the interior of Africa and other parts which are far less favorable to human life than the icy regions of the north.

But with respect to expenditure, I may perhaps be allowed, as I have alluded to that topic, again to call to your Lordship's attention that the 'Resolute' is ready equipped for Arctic service by the munificence of another nation, and to add that other Arctic ships, equally well fitted for the purpose, are lying useless in Her Majesty's dockyards, along with accumulated Arctic stores brought back by the late expeditions, and therefore long since included in the navy estimates; and which, besides, are available only for Arctic service, and, if sold, would be bought at only nominal prices. In addition to the above sources of supply are those already existing on the Arctic shores, which are now studded with depots of provisions and fuel, left from the last and former expeditions, and fit as ever for use, because of the conservative properties of the climate.

But even were the expenditure greater than can thus reasonably be expected, I submit to your Lordship that this is a case of no ordinary exigency. These 135 men of the 'Erebus' and 'Terror' (or perhaps I should say the greater part of them, since we do not yet know that there are no survivors) have laid down their lives, in the service of their country, as truly as if they had perished by the rifle, the cannon-ball, or the bayonet. Nay more, – by attaining the northern and already-surveyed coast of America, it is clear that they solved the problem which was the object of their labors, or, in the beautiful words of Sir John Richardson, that "they forged the last link of the North-West passage with their lives."[11]

Surely, then, I may plead for such men, that a careful search be made for any possible survivor, that the bones of the dead be sought for and gathered together, that their buried records be unearthed, or recovered from the hands of the Esquimaux, and above all, that their last written words, so precious to their bereaved families and friends, be saved from destruction.[12] A mission so sacred is worthy of a government which has grudged and spared nothing for its heroic soldiers and sailors in other fields of warfare, and will surely be approved by our gracious Queen, who overlooks none of Her loyal subjects suffering and dying for their country's honor.

This final and exhausting search is all I seek in behalf of the first and only martyrs to Arctic discovery in modern times, and it is all I ever intend to ask.

But if, notwithstanding all I have presumed to urge, Her Majesty's Government decline to complete the work they have carried on up to this critical moment, but leave it to private hands to finish, I must then respectfully request that measure of assistance in behalf of my own expedition which I have been led to expect on the authority of Lord Stanley, as communicated to me by Lord Wrottesley, and on that of the First Lord of the Admiralty, as communicated to Colonel Phipps in a letter in my possession.[13]

It is with no desire to avert from myself the sacrifice of my own funds, which I devote without reserve to the object in view, that I plead for a liberal interpretation of those communications, but I owe it to the conscientious and high-minded Arctic officers who have generously offered me their services, that my expedition should be made as efficient as possible, however restricted it may be in extent. The Admiralty, I feel sure, will not deny me what may be necessary for this purpose, since, if I do all I can with my own means, any deficiencies and shortcomings of a private expedition cannot I think be justly laid to my charge.

In conclusion, I would earnestly entreat of Her Majesty's Government, while this subject is still under deliberation, that they would be pleased to obtain the opinions of those persons who, in consequence of their practical knowledge and vast experience, may be considered best qualified to express them in the present emergency. And as it must be in the ranks of those officers who would naturally be selected for command of any final expedition that these qualifications will most assuredly be found, I trust I may be pardoned for directing your Lordship's attention to the names (which I put down in the order of their seniority) of Captains Collinson, Richards, McClintock, Maguire, and Osborn. All these officers have passed winter after winter in Arctic service, have carried out those skilful sledge operations which have added so much to our knowledge of Arctic Geography, and have ever, in the exercise of combined courage and discretion, avoided disaster, and brought home their crews in health and safety.

I commit the prayer of this letter, for the length of which I beg much to apologize, to your Lordship's patient and kind consideration, feeling assured that, however the burden of it may pall upon the ear of some, who apparently judge of it neither by the heart nor by the head, you will not on that, or on any light ground, hastily dismiss it. Rather may you be impelled to feel that the shortest and surest way to set the importunate question at rest, is to submit it to that final investigation which will satisfy the yearnings of surviving relatives and friends, and, what is justly of higher import to your Lordship, the credit and honor of the country.

I have the honour to be, etc.,

JANE FRANKLIN.

The Right Hon. Viscount Palmerston, K.G.

25

To: RICHARD M. MILNES[1]
60 Pall Mall, London, [Wednesday] 25 February 1857
Source: Printed transcript, in T. Wemyss Reid, ed., *The Life, Letters and Friendships of Richard Monckton Milnes First Lord Houghton* (London: Cassell & Company, 1890), 2:15
Editorial comment: Italics and format of the transcription are retained here. [*sic*]: Hawkstone. This is most likely a transcription error.

Lady Franklin to R. M. M.

60, *Pall Mall, Feb.* 25*th*, 1857.

DEAR MR. MILNES, – I trust you will allow me to offer you my grateful acknowledgments for the able support you have given to Mr. Napier's motion.[2] Not one of the many friends I have seen to-day but has assured me (though the assurance was not necessary) that yours was the most effective speech in aid of his representations. I am sure you do not require to be told that I shall not shrink from the alternative now forced upon me, of completing the search by my own resources. But I have suggested to Mr. Napier whether it might not be urged still upon the Government that they have adopted unprecedented responsibility in announcing their adverse decision without having previously summoned to their aid a committee of experienced Arctic officers, best fitted to advise on the subject. Such a council has always hitherto preceded every decision that has been made. If this be denied, it might still be pressed upon them that the aid they have encouraged me to expect should be freely granted. It was *pecuniary* assistance that Lord Stanley of Alderley spoke of to Lord Wrottesley; but this I do not require, since if I can get the loan of one of the Arctic ships now lying in ordinary, and fit for no other purpose, her fittings completed in the dockyard, and some of the stores purchased which are now lying waste, my resources will be relieved as much as by money. Some concession, I think, is due to me in consideration of that cruel suspense in which I have been kept ever since the return of the last great Fish River Expedition,[3] extending for a period which has made me lose three consecutive seasons while still in doubts as to what answer I should get to my repeated applications as to the memorial of my friends. This state of suspense, ending now in fatal certainty, after preliminary circumstances which, a few days ago, seemed to promise quite a different result, has seriously affected my interests in many ways.[4]

I was extremely pleased to see your mention of the *Resolute*, and am sorry that Mr. Napier did not make use of a document I sent him, which has not anywhere met the notice it deserves. I mean, the Resolution of Congress when it voted the money for the purchase and equipment of the ship.[5] The resolution expressly stated that the object of this noble gift was, not the exhibition of national goodwill, but simply of sympathy with the cause to which the ship had been devoted. I am glad to have an opportunity of telling you that Captain Hawkstone, of the *Resolute*, has placed his services in the most spontaneous and unconditional manner entirely at my disposal for the command of a private expedition, in the event of my being driven to this painful alternative, and in case of his not interfering with any English officers.[6] I shall accept his generous offer only in case Sir Charles Wood refuses to fulfill his own written pledge (communicated to me last year by order of Prince Albert) that willing permission should be given to any naval officer volunteering for my private expedition.[7]

I do not know whether I may ask of you to communicate with Mr. Napier upon the possibility of any further movement in the House. I am afraid of appearing to expect too much, and would rather assure you how much I appreciate what has already been done for me.

Yours very truly,

JANE FRANKLIN.

26

To: SIR CHARLES WOOD
60 Pall Mall, London, [Saturday] 4 April 1857
Source: Printed transcript, *HCSP*, 1857–58/LX/46–8
Heading: CORRESPONDENCE respecting Her Majesty's Ship "Resolute," and the Arctic Expedition. | No. 38 | *Lady Franklin to Sir C. Wood.*

Sir, 60, *Pall Mall, April* 4, 1857.

THE decision so unexpectedly announced by you to the House of Commons against completing the search for my husband's expedition, having thrown upon me the responsibility of carrying out that work by my own resources, I should at once have addressed you on the subject, but that I was strongly advised to wait until the agitation of the elections had passed away, and left me a better opportunity of securing your favourable consideration.[1]

The object of my present appeal is (in reliance on the promises already given me by Her Majesty's Government), to procure such moderate aid as may render my own funds more fully available than they can be without it; and I have therefore earnestly to request that you will be pleased to place at my disposal the "Resolute," lately restored to Her Majesty's navy by the Americans, in the condition in which she arrived in England.

I have the less hesitation in making this request, because the "Resolute" will thereby be restored to the destination which it would appear the American Congress / had in view when they framed the Resolution which described the gift "as a token of the deep interest felt in the United States in the service in which that ship was engaged" when abandoned, and when they voted a large sum of money in refitting her for such service, which money might have been spared, had the only object been to secure her safe passage across the Atlantic.[2]

I need not tell you that, in the same spirit, the individual chosen to execute this mission was the most distinguished Arctic officer in the American navy, who had been sent to seek, and had brought home in safety, the missing navigators, Dr. Kane and his party; and that the person into whose hands was committed by his Government the entire re-equipment of the "Resolute," was Mr. Grinnell, of New York, the main promoter of all the American Arctic expeditions. I may, perhaps, be permitted to add, that such a disposal as I plead for this "consecrated ship," as a distinguished American has called her,[3] would be extremely agreeable to the American people, and would be well calculated to remove any unpleasant feeling existing in the United States at seeing this messenger of the national sentiments consigned to the ordinary condition of useless or superfluous vessels – a feeling not the less strong because a just self-respect and a due appreciation of the delicacy of their own position forbid its undisguised expression to the English nation.

The "Resolute," however, is not the only Arctic ship now lying useless in Her Majesty's dockyards; and if there be any reason, of which I am ignorant, why she should not be again employed in the service for which she was refitted in the navy-yard at New York, I would beg to state that I should be grateful for any one of those other five Arctic ships, or for the use, as a tender, of one of Her Majesty's gun-boats of the smallest class, that is not otherwise wanted, with permission given to the very able and experienced officer who does me the favour of taking command of the expedition, to select what he may think necessary for her out of the Arctic stores now lying in the dockyards.

I would draw your attention to the fact that the concession thus pleaded for, involves no expense to the nation, since the objects indicated have already been included in the Estimates; and I would assure you that the assistance I thus venture to claim, in fulfilment of those general promises of pecuniary support which were conveyed to me by Lord Wrottesley, on the authority of one of

Her Majesty's Ministers, is not pressed for by me with the view of sparing the expenditure of my private funds, which will be devoted, without reserve, to the expedition.

But I may be allowed to remind you that, in one shape or another, Government assistance has never been withheld from me on former occasions. In each of the three little expeditions I have sent out during the last eight years, the Admiralty (besides giving free permission to naval officers to serve, and gratuitously supplying me with pemmican), liberally advanced to me large sums on my husband's pay, at one time to the extent of 2,000*l*. On the last occasion they kindly allowed my vessel to be refitted at the dockyard at Woolwich,[4] free of cost, and furnished me with orders for the commanding officers, dockyard officials, &c., in the Pacific, authorizing them to give to my expedition such assistance as might be required, including in this general order permission for seamen to volunteer in exchange from Her Majesty's ships.

It is natural and inevitable I should feel that if, under three successive Administrations, I was thus befriended, and that, too, under circumstances not so strongly engaging public sympathy in my behalf as the present, it is not at this moment that, for the first time, similar kindness should be denied me.

I do not anticipate any such result, yet I cannot refrain from expressing to you what I deeply feel, though I do so with the earnest desire to avoid saying anything that could justly displease you, that, if after the long state of suspense in which I have been kept, and the reasonable hopes I have been allowed to entertain, you were to refuse me all help, except that for which I have your written pledge, conveyed to me last year, by order of His Royal Highness the Prince, I should feel myself not only most hardly, but unjustly dealt with, and such is, and will be, the prevailing public feeling. No one knows but myself, and those who are immediately around me, how cruelly my health has suffered, and all my domestic affairs have been disturbed by this intolerable suspense; but all can understand, that fearful must be the trial, prolonged month after month, and year after year, in which the life we seek to save is jeopardized by delays, and the darkness to be dispelled, thickens with time. No one but must feel that if this harassing period of alternate hopes and fears were to end at last in the rejection of all my own unanswered appeals, / and those of others, such a decision should have been come to long ago, and mercifully communicated to me at the earliest possible period. My own exertions have been effectually paralyzed, while one precious season after another has passed away; and yet I have never concealed the fact, though shrinking as much as any one can do from obtrusive or independent action, that upon the Government decision depended the immediate execution of my own efforts. I was impelled to this course of action by a solemn sense of duty, and am committed to it by repeated pledges; yet, notwithstanding, the whole subject has been dispassionately considered by my friends and myself, since your announcement in the House of Commons. I may say, therefore, that it is not without the support and sanction of those on whose judgment even Her Majesty's Government would themselves most rely, had they deemed it desirable on this, or on former occasions, to take counsel of them, that I am preparing for the search of that very limited area which remains to be explored, guided to it by a clue the certainty of which the Admiralty have themselves amply recognised by their reward to Dr. Rae.

The yet undiscovered fate of the lost navigators has very recently been declared by the Americans at a solemn public meeting, to be *the* question of this age of active enterprise, and it is certain it will never be set at rest so long as the area alluded to remains unexplored.

I am sure you must wish that my expedition should accomplish its object, and return in safety; and no one knows so well as you, that on the efficiency of the means employed, must depend, humanly speaking, those blessed results, and the consequent finality of the subject. Impressed with this conviction, I entreat your kind indulgence of my earnest words, and your generous consideration of my requests.

It is my anxious desire, and that of my advisors, that the expedition should be equipped in time to take advantage of the present singularly favourable season (for such is anticipated) instead of being deferred to the autumn, for the route by Behring's Straits. The latter route has advantages peculiarly its own, but it would entail a longer voyage and a longer absence, besides being more costly. It will not, therefore, be adopted by me, unless it becomes absolutely impossible to equip an expedition for the eastern route before the close of June.

I have, &c.

(Signed) JANE FRANKLIN.

27

To: [WILLIAM HAMILTON]
60 Pall Mall, London, [Tuesday] 21 April 1857
Source: Printed transcript, *HCSP*, 1857–58/LX/48–50
Heading: CORRESPONDENCE respecting Her Majesty's Ship "Resolute," and the Arctic Expedition. | No. 39. | *Lady Franklin to the Secretary of the Admiralty.*
Editorial comment: [*sic*]: even

Sir, 60, *Pall Mall, April* 21, 1857.

IN the letter I had the honour to receive from you on the 8th instant, in reply to one I addressed to Sir Charles Wood, requesting to be allowed the use of Her Majesty's ship "Resolute," or of some other vessel at their Lordships' disposal for Arctic service, I am informed that the Lords Commissioners of the Admiralty are not aware of ever having held out to me any expectation of sending or of aiding in any expedition since they came to the conclusion that there was no prospect of saving life by any further search in the Polar Regions.[1]

This decision was announced to the House of Commons, in reply to the motion made by Mr. Napier on the 24th of February last, up to which time, or within a few days of it, I had every reason to believe that the deliberations of Her Majesty's Government would have had a totally different result.

But I allude to the date of their Lordships' decision only because you are desired to acquaint me that my Lords are not aware of having even held out to me any expectations of the nature alluded to "since" they came to the above painful, and to me, unexpected conclusion.

I feel great difficulty in treating this point, since it would appear as if I had been guilty of the extreme folly of affixing the promises of assistance held out to me by Her Majesty's Government to the short period which has elapsed since an adverse decision was come to. But as I know that this cannot have been their Lordships' meaning, and feel that it would be disrespectful to affect to treat it as such, I am reluctantly compelled to adopt the only alternative which presents itself to me, and to conclude that the expression made use of is intended to convey to me their Lordships' desire to ignore or withdraw from previous promises of help for a / private expedition, which should not be confounded with a national one, but which were made to me with the express object of meeting the contingency that now presents itself, viz., that of the Government objecting to continue the search, and of my consequently undertaking this labour on my own resources. For I must again take the liberty of reminding their Lordships that it was with express reference to a private expedition that Lord Stanley of Alderley (who was the organ of the Government in the House of Lords upon this question at the close of the last session), assured Lord Wrottesley, who brought the question forward, of being very favourably disposed towards a proposition Lord Wrottesley then made to him, and which was, "that pecuniary aid should be rendered to me for a private expedition next (that is to say, the present) spring," in case the serious consideration of the whole subject, to which Her Majesty's Government had just publicly pledged themselves during the recess, should result in a decision unfavourable to a Government expedition.[2]

This communication was so distinct and positive that Lord Wrottesley, who had given me no previous insight into his intentions, and to whom I had never made any request or suggestion, did not hesitate to communicate it to me immediately by letter. But I allude to the fact, at present, merely to justify the deductions in which I have since implicitly confided, and not with any hope that an explanation of some of the causes by which I have been misled, will induce their Lordships to respect facts which they probably do not see fit to recognize, or for which they do not desire to be responsible.

But the case is different with the pledge I have received from Sir Charles Wood through the intervention of His Royal Highness Prince Albert, and which was forwarded to me last year by His Royal Highness's commands. In that letter Sir Charles Wood, while expressing the objections the Admiralty then entertained, as they do now, against any further search (the circumstances being precisely the same), yet declares his entire readiness to give leave to any naval officer who desires to volunteer for my private expedition; and adds, that the Admiralty will also make and furnish me with pemmican, if I desire it, but refers me for ordnance stores to the proper Department.

It is in reference to this official promise, which I had the satisfaction to observe was not confounded in your official letter with other promises not recognized, that I venture to make an observation, and request a favour.

The officers who have severally done me the favour and the honour of offering to command my expedition, have all attained the highest rank by promotion in the service, and have nothing to gain, in that respect, from the highest degree of success which may attend their exertions in behalf of the claims of humanity. This fact makes me only the more anxious that the kindness of the Admiralty should be extended to such disinterested devotion, and to hope that if my Lords cannot distinctly approve of their volunteering on this occasion, they will, at least, in granting leave of absence, generously forbear from the expression of any discouraging sentiment.

I need not remark, in forwarding with this letter one which Captain M'Clintock has entrusted to my care, that, in the character of that officer, their Lordships will perceive a guarantee for all that ability and great experience and discretion, combined with courage, can effect. Captain M'Clintock has deliberately considered the object to be attained, and the means at his disposal, and has no doubt of the result. Had any expedition taken the western route, it would have been commanded by an officer scarcely less distinguished, in whose ability and zeal I could have placed unlimited confidence.

I think it right their Lordships should be informed that in consequence of their declining to grant the loan of the "Resolute," or of any other vessel, I was under the necessity of taking instant measures to secure a fit substitute, which I have found in a new diagonally-built yacht now lying at Aberdeen, which Captain M'Clintock considers admirably fitted for the purpose, and which the builder engages to strengthen and have ready for sea in six weeks from this time.

In making this communication, I do not hesitate to acknowledge that I am aware, by letters received from America this morning, of the arrival of a representation from New York respecting the "Resolute," which induces me yet to pause a day or two before the purchase of the yacht is irrevocably concluded.[3] I have not seen the document in question, nor have its contents been communicated to me, so that I should perhaps apologise for alluding to it, were I not aware that its tendency must be to support my well-known / wishes, which are also those of the American community. Should this expression of a feeling on the part of a portion of that generous nation which purchased and restored Her Majesty's ship to this country, not be deemed of sufficient importance in the eyes of the Lords Commissioners of the Admiralty, it will still help, I trust, to convince their Lordships that I have not misrepresented American feeling on the subject. Their Lordships may judge what the sentiment must be which could lead an officer of the high standing of Captain Hartstene, to whom the highest missions in his country's service are open, to express to me, in the most generous and fervent terms, his readiness at any time to command the private expedition I might be compelled to resolve on, provided only he did not interfere with any English officer. It is a source of great comfort to me, in the absence of that sanction from their Lordships which I should have so highly valued, to find the same kind feelings and substantial support on every side. Without any effort of my own, my funds have already received some considerable additions, and these contributions, with the whole of my capital, are now being transferred into the hands of trustees, for the purpose of providing against any possible contingency, and so that the final search may yet be made, even should I not live, viewing the uncertainty of human life, to see the result.[4]

I trust I am not more reckless of human life than are their Lordships, but the same men who volunteer to risk it in Arctic service, would, or might if not so employed, be required to risk it to a greater extent in other services, since it has been well ascertained that the balance of health and safety is in favour of the former; and should the returns of comparative mortality in different quarters of the globe be called for in the coming session, such a result will, I believe, be proved beyond dispute.

There are, I would submit, other considerations in the highest interests of humanity to be put against the calculations of the chances of life or death. Were it not so, Her Majesty's Government would not sanction various enterprises now in progress for the advancement of science, the increase of general knowledge, and the welfare of mankind.

I have, &c.

(Signed) JANE FRANKLIN.

28

To: BENJAMIN DISRAELI
13 Park Place,[1] London, [Saturday] 31 December 1859
Source: Original, H, A/IV/N/10

13. Park Place
St James's
Decr 31. 1859

Dear Mr Disraeli

The accompanying volume was intended for your dear sister[2] as an affectionate & grateful acknowledgment on my part of her kind feelings in seeking to renew our ancient friendship.[3]

On this ground alone you would perhaps allow me to transmit it to yourself, but I am glad also of an opportunity of making it testify to my grateful remembrance of the kind advocacy which in past times you have never refused me when my efforts for Arctic search needed support in Parliament.

Believe me dear Mr Disraeli
very truly yours
Jane Franklin.

29

To: BENJAMIN DISRAELI
13 Park Place, London, [Saturday] 3 March 1860
Source: Original, H, A/IV/N/13
Editorial comment: Mary Anne Disraeli has written on the reverse of the letter '1860. March 3$^{\underline{d}}$ | Lady Franklin.'

Dear Mr Disraeli

Allow me to say in few words how deeply I feel the most beautiful & kind things you said last night in the House in supporting Sir F. Baring's motion.[1] Such words might well make me proud, and will assuredly always make me grateful.

yours very sincerely
Jane Franklin

13. Park Place
St. James's
March 3. 1860.

AFTERWORD

After 1860, British interest in Arctic navigation became more of a nostalgic fancy than anything resembling the obsession that had gripped England between 1818 and 1854. When the mystery of Franklin's fate was finally uncovered in 1859, the press, Lady Franklin, the government, and the public rightly and collectively acknowledged that the enterprise of Arctic exploration could – and should – be honourably closed. A memorial to Franklin, and thus also to Lady Franklin, was constructed in Waterloo Place; an obelisk was erected on Beechey Island (the site of the Franklin expedition's first winter quarters); a commemorative plaque was placed in Westminster Abbey; and memorials were designed and erected around the country in tribute to the 129 lost adventurers. These memorials still exist, but their significance has diminished in the wake of the last 150 years of history.

Lady Franklin's "Battery," that suite of rooms in Spring Gardens from which she launched so many well-aimed attacks against the Admiralty, now houses a theatre, the original building having caught fire in the 1920s and been replaced in art deco fashion. Franklin, still standing in bronze in Waterloo Place, now looks over a parking lot. Opposite him stands Robert Falcon Scott, that perhaps more famous polar explorer who met his unfortunate end at the other – though equally vicious – pole. Although Franklin remains the most famous northern explorer, Scott is perhaps the better-known leader of polar disasters because his diaries – and some members of his expedition – survived. Yet, while Scott's first-hand account makes compelling and emotionally gripping reading, it does not diminish the drama of Franklin's own disaster on the south shore of King William Island.

The absence of a written record of the Franklin expedition's experience has left the field open to several imaginative writers inclined to fill in the space. John Wilson, for example, with *North with Franklin: The*

Lost Journals of James Fitzjames (1999), and David Solway, with *Franklin's Passage* (2003), imagine the sequence of events that led to the expedition's destruction. As Wilson articulates in his introduction, those interested in Franklin (or in any of his men) inevitably ask the question, "What was going through his mind?" and in asking, perpetuate the continual imaginative recreation of his expedition's fate. The gap remains, however, between fact and fiction. What Franklin's expedition failed to give us – a written and personal account of the tragedy – I believe Lady Franklin's letters can do much to replace.

As it did for Wilson and Solway, the dearth of first-hand accounts from the Franklin expedition perhaps gives us a feeling of personal connection with Franklin and his men, since we as readers or historians are free in large part to imagine the explorers' experiences as they moved from enthusiasm and early triumph to exhaustion, fear, and death. Indeed, my own continuing interest in the Franklin expedition began with just such an imaginative connection. My first memory about the expedition is from adolescence, sitting cross-legged in my parents' living room, riveted by colour photographs in *National Geographic* showing Owen Beattie's 1984 exhumation of John Torrington from his grave on Beechey Island. Looking at the face of a man not much older than I was, whose clothes still bore the embroidered initials of his brother (an intimate detail that reminded me of my own fragility), I wondered how this sailor could have travelled so far from home and how his life could have ended in such a place. That wonder, shared no doubt by many readers, was increased 129 times when I took in the size of the whole expedition. And my sense of the tragedy was further compounded when I thought of the family members and friends who had been left behind in England, to worry and to imagine.

Like the explorers she strove to save, Lady Franklin herself was an adventurer. She ran up against insurmountable odds – the Admiralty, Victorian attitudes about women, the fickle and occasionally vicious press – in her quest to discover her own passage to her husband. Like her husband's ships, her own expedition failed in its original purpose, but what it discovered, and what it came to mean to England and to the history of exploration, became so much more than success could ever have effected. In her letters and her character, Lady Franklin exhibited the traits necessary for true heroism – a ready approach to hardship and a disdain for turning back no matter what the risks. Recalling her husband's description of the ideal English sailor in 1818, Lady Franklin, too, was "ready and thoughtless" in her pursuit of success. Part of what I find so appealing and compelling about Lady Franklin, as evidenced in her letters, is that even though she was as much in the dark as her contemporaries about what went on in the Arctic, she continued to believe in her husband – if not in his life, then in his honour. Like the small initials on John Torrington's cuff, Lady Franklin's faith in her missing husband showed not only that in great adventure there is always vulnerability, but also that from fragility of the heart can come great strength.

NOTES

INTRODUCTION

1 F.L. McClintock, *The Voyage of the 'Fox,'* 274.
2 Ibid., 262.
3 Parkinson, "'All Well,'" 45. Sir John Franklin's wife, Jane (née Griffin), will be referred to throughout this book as Lady Franklin, in keeping with Victorian titular conventions. Although she was married to a man given a knighthood, she was not a member of the aristocracy and thus would not have been addressed by her first name. This introduction does not contain a biography of Lady Franklin, as her life is not the focus of this collection. For a comprehensive look at her fascinating life, see Ken McGoogan's *Lady Franklin's Revenge*. For a briefer but no less interesting overview, see Dorothy Middleton's text in DNB *Online*: <10089>. Frances Woodward's *Portrait of Jane* offers a somewhat tamer picture of Lady Franklin's life and personality.
4 David C. Woodman has done extensive research on Inuit narratives of the Franklin expedition, narratives that occasionally give much different accounts of the expedition's end. However, the publication of these accounts is relatively recent, and since 1860 the standard narratives of Franklin's fate have followed Lady Franklin's heroic template. See Woodman, *Unravelling the Franklin Mystery*.
5 Berton, *The Arctic Grail*, 318.
6 McGoogan devotes a large portion of *Lady Franklin's Revenge*, his biography of Lady Franklin, to the argument that Lady Franklin's quest on behalf of her husband was "driven by [her] guilt" for having urged him to accept the post, and that "without her behind-the-scenes machinations," Franklin would never have been offered the captaincy in the first place (293). While her ambition for her husband likely played a role in Franklin's selection, I would posit that genuine affection and anxiety also contributed to her desire to discover her husband's ultimate fate.
7 Pierre Berton alludes to the media tempest surrounding the "Franklin Question" during this period as "Arctic fever." Books, dioramas, editorials, and letters circulated in the public sphere, all asking the question, "Where was

Franklin?" (168). A survey of letters, editorials, and advertisements published in the *Times* is a realistic indication of the popularity of the subject. In 1851, for example, the search for Franklin was mentioned thirty-two times in the *Parliamentary Proceedings*, letters to the editor, leading articles, and transcripts of captain's instructions. That said, however, the subject did not much occupy the attention of Parliament. Ian Stone records that between 1848 and 1863 there are only fifty-four references to Franklin in 54,000 pages of parliamentary records. See Stone, "The Franklin Search in Parliament."

8 Both Woodward and McGoogan describe Lady Franklin as "introverted, and painfully shy" (McGoogan, *Lady Franklin's Revenge,* 9). McGoogan complicates this portrait of Lady Franklin, however, by going on to claim that "[a]way from the limelight, which she knew to be dangerous to a woman, Jane secretly thrived on the adulation" resulting from her involvement in the searches (293). Indeed, in spite of her innate shyness, Lady Franklin maintained a forceful public persona throughout and after the period when search parties were sent after the missing expedition, continuing to enter political debates and travelling extensively.

9 See the *Times,* 14 June 1849.

10 *Times,* 13 June 1849.

11 *Athenaeum* 1528 (1857): 177. The letter referred to is Lady Franklin's 2 December 1856 letter to Lord Palmerston (document 24).

12 Brake, "Writing, Cultural Production, and the Periodical Press in the Nineteenth Century," 59.

13 Ibid., 59. The *Quarterly Review* in fact played a pivotal role in reopening the Arctic as a subject of national and scientific interest in the nineteenth century. Sir John Barrow, second secretary of the Admiralty from 1804 to 1845, published a series of articles in 1817 outlining how important it would be to England to discover the Northwest Passage, thereby beginning the thirty-year effort by the Admiralty to sail through the far north. Sir John Franklin's three Arctic captaincies were approved and, in part, planned by Barrow.

14 C. Christ, "'The Hero as Man of Letters,'" 31.

15 Loomis, "The Arctic Sublime," 105.

16 In another context, McGoogan acknowledges Lady Franklin's powers of rhetoric as well as her ability, in a variety of circumstances, to "put on record precisely what [her audience wishes] to hear" (*Lady Franklin's Revenge*, 187).

17 A. McClintock, *Imperial Leather*, 6.

18 Quoted in Harries, "Out in Left Field," 468.

19 See Francis Spufford, *I May Be Some Time: Ice and the English Imagination*, and Robert G. David, *The Arctic in the British Imagination, 1818–1914*.

20 *Athenaeum* 1048 (1847): 1220.

21 Members of Lady Franklin's family viewed her determination to discover her husband's fate as "extreme and possibly ruinous"; her father revised his will in 1851, disinheriting her; and her use of family money to finance searches caused a significant rift between her and Franklin's own daughter from his first marriage (McGoogan, *Lady Franklin's Revenge,* 293).

22 Dr Richard King had participated in an overland Arctic expedition from 1833 to 1835 and had opposed Franklin's 1845 expedition from the beginning. When it was acknowledged that Franklin was most likely in trouble, King strongly advocated an overland search for him. Initially, he was the only person to push for a search in the right area, and had the Admiralty acted on his advice, some survivors from the expedition would likely have been found. Lady Franklin eventually advocated searching in the same area as King. See Wallace, *The Navy, the Company, and Richard King.*

23 The different "dialects" employed by Arctic explorers of the period highlighted the fact that they were writing for a broad spectrum of interested readers, although they generally kept their "scientific" narratives largely separate from their "artistic" ones. See Behrisch, "'Far as the eye can reach.'"

24 Woodward, *Portrait of Jane,* 271. Her suite of apartments in Spring Gardens was affectionately dubbed "the Battery" for the frequent epistolary salvos she would launch from it, but McGoogan also identifies her estrangement from her family as a reason she may have taken the rooms (292).

25 *Fraser's Magazine* 49 (1854): 633.

26 Harries, "Out in Left Field," 467.

27 McGoogan identifies Lady Franklin as a skilled chess player, citing the "clandestine machinations" of social and political negotiation as her "favourite game" (*Lady Franklin's Revenge,* 175). Lady Franklin's position was more than mere rhetoric, however. For Lady Franklin, the personal and the political were inextricably bound, and to ignore one at the expense of the other was to do justice to neither and to negate the power of her claim.

28 Edwin Jesse De Haven and Elisha Kent Kane became American heroes through their involvement with the searches; William Kennedy, a Canadian Hudson's Bay man, was one of Lady Franklin's personal choices as captain for her private expeditions; and Joseph René Bellot, a young French lieutenant, encapsulated the romance of the early searches in his devotion to Lady Franklin. Bellot, a personal favourite of Lady Franklin's, died while in the North on a searching expedition. See Woodward, *Portrait of Jane,* 284–5.

29 Grinnell developed an enduring relationship with the far north, sponsoring a second American expedition (under Elisha Kent Kane) in 1853–55. Sylvia

Grinnell Park, a chief tourist attraction in Iqaluit, Nunavut, is named after his daughter.

30 John Clayton to Lady Franklin, 25 April 1849, H A/IV/N/5b; United States Resolution of Congress, 24 June 1856, *HCSP* 1857–58/LX/26. America's involvement in the Franklin searches spawned a cadre of American Arctic heroes, notably Elisha Kent Kane, surgeon aboard De Haven's expedition (1850–51) and captain of his own expedition (1853–55), both sponsored by Henry Grinnell.

31 Benjamin Disraeli, speech in Parliament, quoted in the *Times*, 13 June 1849, 3b.

32 Spufford, *I May Be Some Time*, 94.

33 *HCSP* 1850/XXXV/316.

34 There is a small series of recommendation letters from Lady Franklin to the Admiralty at *HCSP* 1850/XXXV/316.

35 Goodsir, *An Arctic Voyage*, viii.

36 *Athenaeum* 1678 (1859): 843.

37 See Glyndwr Williams, *The British Search for the Northwest Passage*. Williams offers an interesting and comprehensive look at the impetus, dangers, and rewards awaiting the explorers, both public and private, who took up the challenge of northern exploration in the eighteenth century.

38 Officers generally were promoted when those above them died; in peacetime this was much less likely to happen with regularity. See Fleming, *Barrow's Boys*.

39 Fleming, *Barrow's Boys*, 35.

40 The Arctic was not the only arcane landscape British naval officers found themselves exploring during this interwar period; Africa and its great rivers were also targets for curiosity. Some officers, like George Lyon, went to both Africa and the Arctic in their effort to sustain their employment with the Admiralty. Explorer-observers contributed significantly to Britain's knowledge of the world. John Richardson (the surgeon on Franklin's first and second expeditions), for instance, produced reports on the flora of the far north that were far more comprehensive than any natural histories produced for other parts of North America. See Fleming, *Barrow's Boys*.

41 Cooke and Holland, *Exploration*, 142–3.

42 When nineteenth-century explorers were publishing their journals, the practice in the publishing field was for the publisher to offer the author a set advance, and itself either generate a profit or sustain a loss. Given the popularity of Arctic travel journals, the publisher Murray ran virtually no risk of taking a loss. See I.S. MacLaren, "From Exploration to Publication."

43 Dickens, "The Last Leaves of a Sorrowful Book," 319.

44 Franklin, *Narrative of a Voyage*, 6.

45 Barrow, *Chronological History*, 378.

46 Bloom, *Gender on Ice*, 12.

47 Morley, "Unspotted Snow," 246.

48 Frances Woodward records Lady Franklin's purchase of a monkey for her husband's 1845 expedition (*Portrait of Jane*, 253). In his "Theatre in the North," Patrick O'Neill talks about the popularity of shipboard dramatics in nineteenth-century voyages. See also I.S. MacLaren's "The Poetry of the 'New Georgia Gazette' or 'Winter Chronicle' 1819–1820."

49 Morley, "Unspotted Snow," 246.

50 Episodes of terror served to make the expeditions all the more romantic in the eyes of the audience. For example, the icebergs that each of George Lyon's crew members struggled "to sketch ... as head-pieces for his letters to England" grew in grandeur and significance the worse the 1825 expedition fared (Lyon, *Brief Narrative*, 17). The fact that Englishmen could appreciate the aesthetic beauty of their inanimate foes was further proof of their enlightened state of mind.

51 Morley, "Unspotted Snow," 241. Edward Dramin (1996) makes explicit the link between the mythic and the historical hero: Ulysses embodies a "prominent heroic paradigm of the Victorian era" that was clearly reflected in Franklin's two published narratives and in his public reputation (117).

52 Wiebe, *Playing Dead*, 43.

53 Ibid., 43.

54 McGoogan details Lady Franklin's campaign for appropriate memorials, which lasted long after the fate of Sir John and his crew had become known. See McGoogan, *Lady Franklin's Revenge*, 386–90.

55 For instance, in 1833 Sir John Ross led his crew to safety after having been stuck in the ice for three years and assumed dead. Ross's expedition, launched in 1829, was supposed to last only one year. He and his fellow adventurers arrived home aboard a whaler with all but one of the expedition's members and the discovery of the magnetic north pole in their collective pocket. See Berton, *Arctic Grail*, 107–21.

56 Berton, *Arctic Grail*, 145.

57 Thomas Blanky to his wife, quoted in *Quarterly Review* 92 (1852–53): 388.

58 *HCSP* 1847–48/XLI/188.

59 Quoted in Gibson, "Sir John Franklin's Last Voyage," 10. For a detailed discussion of the food tins aboard Franklin's expedition and their manufacture, see Beattie and Geiger, *Frozen in Time*, chap. 13. Beattie identifies the ingestion of lead as a major factor that compromised the health of Franklin's men (and particularly his officers). Lead was used not only in the soldering process for cans, but also in the general "storage and serving of food," from lead-glazed tableware, pewter ware, and glazed containers to tea and chocolate stored

in "containers lined with lead foil" ("Elevated Bone Lead Levels," 145, 146). Roger Amy and his co-authors (see Amy et al., "The Last Franklin Expedition") conducted a formal autopsy of one of Franklin's crew members, analysing tissue and bone samples for potential cause of death. While they acknowledge that the "significance of the elevated lead levels in determining the course of the expedition remains uncertain," they note that the levels in analysed tissue samples are significantly higher than normal. See also Scott Cookman, *Ice Blink*.

60 Deposition of Captain Martin, quoted in the *Times*, 3 January 1852, 5e.

61 Pierre Berton draws attention to the conflicting ambitions of many Arctic explorers of this period. Empathy and altruism may have originally motivated them to participate, but the attraction of geographical fame was often too hard to resist. Berton's interpretation of McClure's 1850–54 expedition might be applied to others as well: McClure's captaincy of the *Investigator* "was his first command, and he intended to make the most of it. If he could find Franklin, or discover the North West Passage – or both! – he would be the most famous man in England, as well as the richest naval officer of his day" (Berton, *Arctic Grail*, 211). McGoogan goes further, arguing that the Admiralty may in fact have used the "politically popular Franklin search as a cover for seeking the fabled Polar Sea, or at least a way to reach the North Pole" (McGoogan, *Lady Franklin's Revenge*, 357). Sadly, the historical record supports Berton's and McGoogan's suspicions. After Franklin's path was identified as "south," the Admiralty noticeably "lost motivation" for the search (McGoogan, *Lady Franklin's Revenge*, 357).

62 There is ample evidence of search participants willingly giving up payment or security to join the search. William Penny, who captained Lady Franklin's ship the *Prince Albert* in 1850, refused any payment for his time spent up north. Sir John Richardson, one of Franklin's old friends and the surgeon on Franklin's first Arctic captaincy (1819–22), "forfeited a life-insurance policy of £500" in order to search for his friend (Woodward, *Portrait of Jane*, 280). Many naval captains also refused to accept payment from Lady Franklin, but eventually they either relented or were paid by the Admiralty for their services.

63 Quoted in Woodward, *Portrait of Jane*, 285. Eleanor Gell (Lady Franklin's stepdaughter) took this abrupt wardrobe change as "evidence of madness" and used it in her fight to stop Lady Franklin from spending her fortune looking for Sir John (McGoogan, *Lady Franklin's Revenge*, 331). McGoogan identifies this move on Lady Franklin's part, perhaps a classic chess manoeuvre, as "evocative brilliance" (331).

64 Berton, *Arctic Grail*, 264.

65 Belcher was brought before a court martial for his loss of the four ships but was eventually acquitted. See Berton, *Arctic Grail*, 269.

66 John Rae to Lords of the Admiralty, 29 July 1854. Rae was shocked to find his report, written privately to the Admiralty, published in the *Times*. Indeed, the letter he originally wrote to the *Times* did not mention cannibalism. See Richards, *Dr John Rae*, and McGoogan, *Fatal Passage*.

67 See Woodward, *Portrait of Jane*, 285; and McGoogan, *Lady Franklin's Revenge*, 319–27.

68 Dickens, "The Lost Arctic Voyagers," 365.

69 Ibid., 362.

70 Ibid.

71 To refute the implications of Rae's tale, Dickens in fact did much historical research on the instances of shipboard cannibalism in European history, drawing – strangely – on examples of common sailors being unable to withstand the lure of cannibalism. See Behrisch, "On the Trail of an Arctic Tale."

72 I identify Rae as the prophetic Nurse Esther in *The Frozen Deep* for two reasons: Rae and the nursemaid share a Scottish cultural heritage; they both exist outside of the accepted English drawing-room culture (of the real and fictional communities respectively); and (this last is Dickens's argument) they both proclaim stories as factual with insufficient proof. That Nurse Esther in *The Frozen Deep* is proved wrong is no coincidence. See Behrisch, "On the Trail of an Arctic Tale."

73 See C. Stuart Houston, "John Rae (1813–1893)."

74 Woodward, *Portrait of Jane*, 289.

75 Morley, "Unspotted Snow," 242.

76 Berton goes into detail about this. There is also evidence in HCSP that McClure's focus during his expedition was certainly geographical rather than rescue. See Berton, *Arctic Grail*, 219–23.

77 While the march in a southerly direction allowed Franklin's men to accomplish – according to the terms the Admiralty accepted for McClure – the Northwest Passage, this was the direction least likely to lead to rescue. Had they gone north or east, the men would have encountered caches from past expeditions (Lady Franklin mentions these in document 24) or might have been able to contact whalers; the remaining officers of the expedition would have been aware of both possibilities. However, the Franklin expedition had decided to travel south, and there are two possible explanations: they might have been determined to accomplish the Northwest Passage at all costs (a feat for which McClure was also prepared), or the effects of extensive lead poisoning on Franklin's remaining officers could have had a "significant impact on

[their] decision-making processes," leading them to choose a dangerous course (Beattie and Geiger, *Frozen in Time*, 161).

78 *HCSP*, 1854–55/VII/8–9.

79 Ibid., 6.

80 C.R. Markham, *Lands of Silence*, 402. In *Fatal Passage*, McGoogan credits John Rae with being the first to discover the passage while surveying for the Hudson's Bay Company in 1851. Again, the waning interest on the part of the Admiralty supports McGoogan's suspicion that the Admiralty had been using the Franklin search as a façade to justify further exploration of the northern archipelago, having limited interest in the southern Arctic.

81 McClintock was generous in his assessment of the Admiralty's involvement in Lady Franklin's last expedition, reporting in his narrative that "[t]he Government, although declining to send out an expedition, yet now contributed liberally to our supplies" (F.L. McClintock, *The Voyage of the 'Fox,'* 8). These contributions included shot and powder, pemmican, ice gear, lemon juice, and hydrographical and meteorological instruments (8–9). Upon his return, McClintock was credited the time he had served aboard Lady Franklin's ship as "time served by a captain in command of one of Her Majesty's ships" (quoted in F.L. McClintock, *The Voyage of the 'Fox,'* vi).

82 *Fraser's Magazine* 49 (1854): 633.

83 *Times* (London), 15 April 1857, 6f; 30 April 1857, 9a. In "An Episode in the Franklin Search: The *Prince Albert* Expedition, 1850," Ian Stone calls into question Lady Franklin's motives and methods, arguing that while of "her merits there is little doubt, … the extent to which she was honest, scrupulous or principled [in her pursuit of success] affords considerable scope for conjecture" (129).

84 A complete list of contributors appeared in F. Leopold McClintock's narrative of the expedition. See F.L. McClintock, *The Voyage of the 'Fox,'* Appendix 5.

85 F.L. McClintock, *The Voyage of the 'Fox,'* 6–7.

86 Ibid., 11.

87 Ibid., 318.

88 *Leisure Hour* 417 (1859): 816.

89 *Fraser's Magazine* 61 (1860): 221–7.

90 This evidence also essentially corroborated the Inuit story Rae had brought back to England in 1854.

91 *DNB*, 2:631.

92 *Athenaeum* 1665 (1859): 398.

93 Berton, *Arctic Grail*, 142; Woodward, *Portrait of Jane*, *passim.*

94 F.L. McClintock, *The Voyage of the 'Fox,'* 12.

es, [and] a piece of rope with the Woolwich naval mark on it" (*HCSP* 1851/
XIII/268).
ry Grinnell (1799–1874), American philanthropist and shipping magnate
B, 8:2). Grinnell had retired from his own shipping firm in 1850,
drawing completely from the business world, though contributing liberally
aritable causes. Perhaps because of his shipping connections, Grinnell
tained an interest in exploration matters, eventually bearing "the entire
nse" of the first American search expedition. Beginning in 1850, Grinnell
Lady Franklin maintained a correspondence.
May 1850 resolution of Congress can be found in full at *HCSP* 1851/
III/199–200.
ohn Ross (1777–1856), rear-admiral and Arctic explorer, uncle to James
Ross (*DNB*, 17:267–9). Ross was the commander of the first nineteenth-
ury Arctic expedition (1819–20) to search for a Northwest Passage, during
h he misidentified a range of mountains (they were a mirage) and fell per-
ntly out of favour with the Admiralty establishment. During his 1829–33
dition (to which he contributed £3,000), Ross managed to keep his crew
vely healthy and eventually led them to safety after they were forced to
lon their ship. He was knighted in 1834 and given the gold medal of the
raphical Societies of London and Paris. Ross had offered his services in
o command an official Admiralty search for Franklin but was refused. In
he raised funds through public subscription and through the Hudson's
ompany to search for Franklin.
oss, *Polar Pioneers*, 323.
rds, *Dr John Rae*, 81.
Jesse De Haven (1816–1865), naval officer and explorer (*DAB*, 5:200–1).
ven was a much-travelled officer who sailed during his employment with
Navy to the West Indies, the Pacific, Antarctica, and western North
ca. He was chosen for command of Grinnell's US expedition because
olar experience and his "reputation as a scientific officer" (200). De
's naval career ended in 1857, "as his eyesight was affected and his health
ed by his Arctic experiences" (201). For more detail on De Haven's
see Cruwys, "Edwin Jesse De Haven."
:201.
Arctic Expeditions, 537.
hard Collinson (1811–1883), admiral and surveyor (*DNB*, 4:839–40).
on was appointed to the command of the *Enterprise* in 1849, but the
ion did not set out until January 1850. The *Enterprise* and *Investigator*,
Commander (later Sir) McClure, separated "in the neighbourhood of
orn and did not meet again" (839). Collinson's captaincy is admirable:

1 See, for instance, Richard Julius Cyriax's *Sir John Franklin's Last Arctic Expedition* (1939), George Malcolm Thomson's *The North-West Passage* (1975), Roderic Owen's *The Fate of Franklin* (1978), and Noel Wright's *Quest for Franklin* (1959), in addition to the books on Lady Franklin herself: Frances Woodward's *Portrait of Jane: A Life of Lady Franklin* (1951), W.F. Rawnsley's *The Life, Diaries and Correspondence of Jane Lady Franklin* (1923), and Ken McGoogan's *Lady Franklin's Revenge* (2005).

2 I would be unoriginal if I were to say that Lady Franklin is an exceptionally talented writer. Every historian or chronicler who cites her letters expresses this sentiment in his or her own fashion. The skill so evident in the letters themselves makes it unnecessary to stretch them over a theoretical frame. Thus, in the Introduction, Chronology of Events, and the endnotes to each letter, I provide historical context and significant background to the topic at hand, but leave the specific theoretical perspective up to each individual reader.

3 Berton, *Arctic Grail*, 201.

4 "D'Israeli" is the traditional spelling of Benjamin Disraeli's name. Disraeli's first known signature without the apostrophe is on a 1822 document (BDL I/4). Because Lady Franklin was friends with Disraeli's family, I use the original family name in this instance. Only after Isaac D'Israeli's death (Benjamin Disraeli's father) did Benjamin's name begin to be spelled without the apostrophe.

5 Woodward, *Portrait of Jane*, 289.

6 McGoogan, *Lady Franklin's Revenge*, 191.

7 Ibid., 57.

8 *DNB* 2: 613; Roderick Murchison to Benjamin Disraeli, 2 March 1860 [H A/IV/N/12]. In *Lady Franklin's Revenge*, McGoogan goes into detail about the true extent of Lady Franklin's adventures in a useful and enlightening appendix that tallies her travels (425–8).

9 Admiralty Notice, *HCSP* 1854/XLII/339; Lady Franklin to Benjamin Disraeli, 22 March 1849 (document 2); Lady Franklin to Lord Palmerston, 26 June 1855 (document 19).

CHRONOLOGY OF EVENTS, 1845–1859

1 *HCSP* 1847–48/XLI/183.

2 William Penny (1809–1892), whaling captain and Arctic explorer (*DCB* 12: 835–6). Penny had been a whaler since the age of 12, when he first accompanied

his father to Davis Strait. He continued making annual whaling expeditions there, and in 1839 was given permission to explore the shores of Baffin Island, with the plan to establish a winter whaling port there. In between conducting his own explorations, Penny heard of Sir John Franklin's plight.

3 Cooke and Holland, *Exploration*, 177.

4 Sir John Richardson (1787–1865, knighted 1846), Arctic explorer, naturalist, and long-time friend of Sir John Franklin (*DNB* 16:1119–21). Richardson had been surgeon and naturalist on Franklin's first Arctic land expedition (1819–22), during which more than half the crew died from starvation, and he again accompanied Franklin on his second – more successful – land expedition (1825–27). Throughout the searches for Franklin, and after 1859, Richardson remained a strong friend to Lady Franklin.

5 John Rae (1813–93), surgeon, Arctic explorer, and a chief factor of the Hudson's Bay Company (*DNB*, 16:594–6). Rae began his career as a northern explorer in 1846, employed by the Hudson's Bay Company to survey parts of the North American Arctic coast, eventually mapping "more miles of North America's unknown northern coastline … than … any other explorer" (Houston, "Dr. John Rae," 78). Rae was personally chosen by John Richardson as second-in-command of the Admiralty's 1847 overland search for Franklin.

6 Sir James Clark Ross (1800–1862), rear-admiral and Arctic and Antarctic explorer (*DNB*, 17:265–6). Ross was aboard the first nineteenth-century Arctic expedition (1819–20), beginning a career of polar exploration that lasted until 1849. In 1831, Ross discovered the magnetic north pole; in 1841, he discovered Mount Erebus and the Ross Ice Shelf in Antarctica. His 1848 expedition was his eighth voyage to the poles, and his last. Though he never went north again, Ross continued to be consulted by the Admiralty during the next decade of the Franklin search.

7 Smith, *Arctic Expeditions*, 482.

8 Sir Henry Kellett (1806–1875), vice-admiral (*DNB*, 10:1230). Kellett was employed by the Admiralty primarily as a southern surveyor; his career began in the West Indies and continued along the coasts of Africa and South America. In 1845, Kellett began his command of the *Herald* as a surveyor along the coast of Colombia. The three summer searches in the western Arctic were temporary interruptions in this six-and-a-half-year voyage.

9 Thomas Edward Laws Moore (1819–1872), rear admiral and governor of Falkland Islands (1855–62) (*MEB*, 2:955). Though he had no Arctic experience, Moore had been a mate on the *Terror* during James Clark Ross's 1839–43 Antarctic expedition, and had commanded the *Pagoda* on an Antarctic expedition in 1845 (M.J. Ross, *Polar Pioneers*, 298).

10 *HCSP* 1849/XXXII/213. W. Gillies Ross argues that there organized council created to direct the Franklin search and that its historical existence as a cohesive body in ch apocryphal. See W. Gillies Ross, "The Arctic Council of

11 Smith, *Arctic Expeditions*, 484.

12 Zachary Taylor (1784–1850), twelfth US president (1849-

13 *Times*, 22 May 1849, 8b.

14 *HCSP* 1850/XXXV/323.

15 Horatio Thomas Austin (1801–1865) (*DCB*, 9:11–13). Aus and successful naval career, serving in the War of 1812 and Latin America and in the English Channel and the first Arctic experience was with Sir Edward Parry (see 1824. Austin was known in the nineteenth century for steam technology, for which he was chosen to comma His 1850 expedition has been described as "one of the expeditions to that date" (13).

16 Osborn, *Stray Leaves from an Arctic Journal*, 30.

17 Sir Erasmus Ommanney (1814–1904), admiral (*DNB*, Ommanney was only thirteen, he took part in the ba later spent most of his naval career in the Mediterran voyage to the Arctic had also been a rescue mission; Captain (later Sir) James Clark Ross, Ommanney hel of whalers caught in the ice in 1833. In 1850, Captain chose Ommanney to be his second-in-command, ha in the Mediterranean. Ommanney was knighted in Franklin's encampment.

18 Ommanney, quoted in Beattie, *Frozen in Time*, 22. tins of preserved meat and pieces of torn clothing.

19 Smith, *Arctic Expeditions*, 516. For more detail abou pute, see Holland, "The Arctic Committee of 1851:

20 Charles Codrington Forsyth (c. 1814–?). Forsyth w Lady Franklin, by whom he was "especially liked," vate expedition (Woodward, *Portrait*, 272). The Fr 1842 in Van Diemen's Land. In 1850, he offered his for free. The expedition brought back the first Fra three miles east of Beechey Island on Cape Riley. Admiralty on his return, he described having see had been pitched, or stones placed, as if they had lower part of the tents down; also great quantities

his men were relatively healthy during a five-year trial in the Arctic and were able to sail their own ship back to England. He also discovered a Northwest Passage, but because he received no official recognition for this or for his overall accomplishment, he never applied for Admiralty employment again.

30 Sir Robert John le Mesurier McClure (1807–1873), vice-admiral and Arctic explorer (*DNB*, 12:448–9). When he set sail in 1850, McClure had already travelled to the Arctic twice, first in 1836 and then in 1848 as a member of Sir James Clark Ross's search expedition. McClure's appointment to the *Investigator* was a mark of his favour with the Admiralty; it was his first command. After losing contact with Collinson in the *Enterprise*, McClure continued into the western Arctic alone, pushing as far as he could into the pack by ship. In 1853, McClure made the decision to abandon his ship, and he and the remainder of his crew were rescued by a sledging crew from the *Resolute*. McClure and his men were the first Englishmen to cross from one side of the Arctic to the other, entering through the west and exiting through the east. He was initially recognized as the first discoverer of a Northwest Passage, and for this was knighted in 1855.

31 McClure, quoted in Smith, *Arctic Expeditions*, 542.

32 William Kennedy (1814–1890), fur trader, explorer, and sailor (*DCB*, 11:470–1). A deeply religious man from a family of traders, Kennedy resigned from the Hudson's Bay Company in 1846 after disagreeing with the company's policy of selling liquor to Natives. Between 1846 and 1851, when he travelled to England to command Lady Franklin's second expedition, Kennedy established a fishery, worked as a boat captain, and lobbied against the HBC's monopoly. Though not an Admiralty choice, Kennedy was well prepared, with his extensive northern trading experience, for an expedition to the Arctic. He chose Canadian traders for his crew and insisted upon Native clothing for the expedition.

33 Sir Edward Augustus Inglefield (1820–1894), admiral, "a man of cultivated taste and mechanical ingenuity" (*DNB*, 22:904–5). Inglefield had seen active service before commanding Lady Franklin's private vessel, serving in Syria (1840) and in South America (1845). His role as an Arctic explorer seems to have been mainly as a go-between. In his first voyage on the *Isabel*, he brought back the news of Ommanney's discovery of an encampment; in his 1853 voyage on the *Phoenix*, he returned with the news of McClure's discovery of a passage.

34 Sir Edward Belcher (1799–1877), admiral and surveyor (*DNB*, 2:142–3). Belcher's first Arctic service was in 1825 aboard the *Blossom* as a surveyor on an expedition that lasted three years. He was knighted in 1843 for other surveying operations, and he published a popular narrative of his seven-year voyage commanding the *Sulphur*. While in other contexts Belcher was a competent commander, his term as senior officer in the 1852 Arctic search was disastrous for his career;

"no officer of equal ability has ever succeeded in inspiring so much personal dislike," and he was never employed by the Admiralty again (143).

35 Sir Francis Leopold McClintock (1819–1907), admiral and Arctic explorer (*DNB*, 2nd suppl., 2:506–7). McClintock's first Arctic voyage was under Sir James Clark Ross in 1848; upon his return to England, he was appointed first lieutenant under Erasmus Ommanney aboard the *Assistance* (in Austin's squadron). On this expedition, McClintock distinguished himself through his experiments with sledge travel, journeying "an unprecedented 760 miles in 80 days" (506).

36 Lady Franklin's 1853 expedition was partially sponsored by colonists of Van Diemen's Land; they contributed £1,871 13*s*. 4*d*. (Barr, "'The Cold of Valparaiso,'" 205). Kennedy had the crew jailed for mutiny, but was unable to find men to replace them because of the impending Crimean war. He returned to England in expectation of another expedition, but when it did not materialize, he returned to Canada in 1856 (*DCB*, 11:471). For more detail on the events leading up to and following the mutiny, see Barr, "'The Cold of Valparaiso.'"

37 Elisha Kent Kane (1820–1857), physician, naval officer, explorer, and "America's first Arctic hero" (*ANB*, 12:368–9; *DAB*, 10:256–7). Kane was an inveterate traveller, "fervidly seeking adventure" in the capacity of assistant surgeon in the US Navy (*DAB*, 10:256). He applied for and was accepted to the post of senior medical officer of Grinnell and De Haven's 1850 Arctic expedition, and on its return, he soon began planning another search for Franklin. Kane's first captaincy was his catalyst to fame; the expedition's abandonment of its ship and trek to the coast of Greenland stand "in the annals of Arctic exploration as an archetype of victory in defeat" (257). See also Corner, *Doctor Kane of the Arctic Seas*.

38 Lieutenant Henry J. Hartstene (d. 1868), United States Navy (NMMC II/ii/693). Hartstene became involved in the Franklin search at its most difficult point, in 1855. Collinson and his men were still stuck in the Arctic; the Admiralty had just lost five of its ships; the Northwest Passage had recently been discovered; Rae had just returned to England with news of the expedition's possible fate (see below); and the American expedition, outfitted through the generosity of the US government and Henry Grinnell, was now considered in trouble. Despite the negative publicity, however, Hartstene remained an enthusiastic supporter of Lady Franklin and her cause. See document 25.

39 Richards, *Dr John Rae*, 101.

40 Rae to Admiralty, printed in the *Times*, 23 October 1854.

41 Rae, quoted in *Household Words* 254:20. Rae's full report of cannibalism was printed in the *Times*, but copies of this letter with the cannibalism detail excised were circulated. The transcriber of one such copy admits to "omitting

only a few minor descriptions" in his/her transcription; only the cannibalism "description" is missing (SPRI, 248/459/1). More significantly, Rae himself provided the *Times* with a letter that did not mention cannibalism; the Admiralty was responsible for the publication of the full letter.

42 *HCSP* 1854–5/VII/7.

43 James Anderson (1812–1867), HBC chief factor and explorer (*DCB*, 9:5–6). Anderson was chosen by the Hudson's Bay Company to lead the small canoe expedition to the Back River. The expedition was delayed in starting, the spring being unusually late, and Anderson tried unsuccessfully to employ an Inuit interpreter before setting out. The brief expedition had a profound effect on Anderson's health; he permanently lost his voice from exposure and eventually died of tuberculosis. For his recovery of artifacts from Montreal Island he received the Polar Medal and £400 from the British Admiralty.

44 James Green Stewart (1825–1881), HBC chief factor and Arctic explorer (*DCB*, 11:854–5). Stewart was given the second-in-command of the HBC canoe expedition, but relations between Stewart and Anderson were strained from the beginning. Anderson blamed the frailty of the boats on Stewart's choice of "poor bark" in their construction, and did not approve of the crew Stewart had originally chosen (854). After the expedition's return, Stewart was reprimanded by the Hudson's Bay Company for his alleged lack of initiative. For his participation in the expedition, he received £280 from the Admiralty.

45 Barr, *Searching for Franklin*, 245. Barr provides extracts of letters and reports about Anderson and Stewart's expedition; see Tyrrell, "A Story of a Franklin Search Expedition." Stories like this circulated in the nineteenth century but remain unverified. In *Beyond the Hills of Dream*, William Wilfred Campbell renders in dramatic monologue just such a confession; in the chapter "Unabsolved" (95–106), he dramatizes "the confession of a man who went with one of the expeditions to save Sir John Franklin's party, and who, being sent ahead, saw signs of them, but, through cowardice, was afraid to tell" (95).

46 *HCSP* 1856/VII/333–7.

47 The £10,000 was one part of the original £20,000 announced in 1850; before Rae, no one had enough information to claim it.

48 *HCSP* 1856/XLI/167.

49 *HLSP* 1856/X/399. Although Rae did receive the reward for obtaining first evidence of the expedition's fate, his discovery most likely cost him a knighthood (an honour bestowed on other searchers for lesser discoveries). See Richards, *Dr John Rae*.

50 Henry John Temple, 3rd Viscount Palmerston (1784–1865) and Tory prime minister from 1855 to 1865 (with the exception of 1858–59, when he resigned over a bill he was unable to pass) (*DNB* online).

51 The ship was eventually broken up. Some of its timbers were made into desks, one of which was presented to Lady Franklin, another of which is now used in the Oval Office in the White House (Vancouver Maritime Museum).

52 W.G. Ross, *This Distant and Unsurveyed Country.*

53 McClintock had returned to England in 1854 after his unsuccessful expedition under Commodore Belcher. When offered the command of Lady Franklin's *Fox* for a final searching expedition, McClintock accepted and offered his services for free, obtaining leave from the Admiralty to sail under Lady Franklin's directions. He was knighted in 1860.

54 F.L. McClintock, *The Voyage of the 'Fox,'* 8–9.

55 Ibid., 39.

56 Ibid., 90.

57 William Robert Hobson (1831–1880), naval captain and explorer (*MEB*, 1:1490). Hobson had been a mate aboard a supply and relief expedition sent to assist the *Enterprise* and *Investigator* in 1853, and later served in the Crimean war. He was in charge of the sledge expedition that uncovered the Franklin record on Victory Point.

58 Parliamentary Minutes quoted in the *Times,* 3 March 1860, 6d.

59 This quotation has been often used in Franklin tributes, both nineteenth-century and contemporary. It is on Franklin's plaque in Westminster Abbey, in Leopold McClintock's book about his discovery of Franklin's fate, and in the 2005 film by Crossing the Line Films, *Finding Franklin.* The quotation epitomizes the sense of endurance and self-sacrifice that characterizes the whole of nineteenth-century Arctic exploration.

THE DOCUMENTS

Document 1

1 Lady Franklin's reward offer to whalers marks the beginning of the search for Sir John Franklin's expedition. Although the Admiralty had made enquiries in 1847 about the possibility that the expedition might be in trouble, naval officers still considered it likely that Franklin and his crew would be home by 1848.

2 Lady Franklin's family residence and the home of Sir John Simpkinson (1781–1851), Lady Franklin's brother-in-law (*POD*, 1850; *BDL*, V/1897n5). Lady Franklin stayed there when Sir John was away on expeditions.

3 Sir Francis Beaufort (1774–1857), rear admiral and hydrographer to the navy, 1829–55. Beaufort was an influential member of the Admiralty and the Royal Society, and lent his support to many scientific enterprises. His name was synonymous with hydrography and nautical science (*DNB*, 2:39–41).

4 Sir William Edward Parry (1790–1855), rear admiral (1852) and Arctic explorer. Parry's first Arctic voyage was in 1818. In 1819 he returned to the north as a captain to look for the Northwest passage, and again in 1821, 1824, and 1827. Although Parry was unsuccessful in his efforts to find a passage, he was known for reaching the farthest west and the farthest north accessible by ship. His record was not broken until 1876 (*DNB*, 15: 392–3; Cooke and Holland, *Exploration*, 239).

5 Thomas Ward, owner of the *Truelove*, a whaling ship that included in its hunting route a search for Sir John Franklin. Hull was a whaling port in northern England.

6 On 6 March, the Admiralty requested notice to be given in whaling ports that "not less a sum than 100 guineas" would be given to "the crew of any whale ship that may bring accurate information of Her Hajesty's [*sic*] 'Erebus' and 'Terror'" (*HCSP*, 1847–48/XLI/209). On 13 March, this offer was augmented with a promise of "higher reward" for verifiable information received (*HCSP*, 1847–48/XLI/210). Their offer of reward was distributed with Lady Franklin's (document 1).

Document 2

1 Benjamin Disraeli (1804–1881), a friend of Lady Franklin's family and leader of the Protectionist Party (*DNB*, 5:1006–22). Consistently outspoken on matters of social and national justice, Disraeli was a perfect parliamentary representative for Lady Franklin in her attempts to motivate the British government to search for her missing husband.

2 Spring Gardens is near the Admiralty building in Whitehall; Lady Franklin kept an office here to meet with Arctic navigators and keep abreast of the latest developments regarding her husband (Woodward, *Portrait of Jane*, 271).

3 John Philip Gell, family friend of the Franklins, fiancé to Franklin's daughter (from his first marriage) Eleanor, and former warden of an experimental college established by Franklin in 1840 while lieutenant-governor of Van Diemen's Land (Woodward, *Portrait of Jane*, 263).

4 The reason for this appointment is undocumented, Disraeli makes no mention of it in his personal correspondence, and there is no relevant discussion in Parliament, but it most likely refers to a revised reward Lady Franklin was about to offer British whalers (see document 3). There are, however, also some other options: about this time Lady Franklin was equipping the first of her privately funded expeditions with the *Prince Albert*, and she was requesting a change in sailing orders for the Admiralty supply ship *North Star* (Woodward, *Portrait of Jane*, 264).

Document 3

1 See document 1n3.
2 See document 1n4.
3 See document 1n5.
4 Significantly, this second notice to whalers lacks any reference to Sir John Franklin as a possible referee, as in the 1848 notice (see document 1). Although Lady Franklin is still confident of her husband's continued survival, her omission of this reference and increase in reward are ominous.

Document 4

1 Lady Franklin visited the United States in the summer of 1846 with Sir John's daughter Eleanor (Woodward, *Portrait of Jane*, 255).
2 Until 1859, there was considerable public confusion over the exact number of men who sailed under Sir John. The original number was 134, but 5 returned to England before the ships entered the Arctic; 129 men went missing. Although supplied with canned food sufficient for three years, the ships were also stocked with ammunition, to preserve provisions when on land (Beattie and Geiger, *Frozen in Time*, 164–6; *HCSP*, 1854/XLII/340–53; F.L. McClintock, *The Voyage of the 'Fox,'* 317).
3 It took up to six months for information to travel from Great Slave Lake to England, so Lady Franklin's claim here is not necessarily the case (though it did prove later to be true). In addition, Richardson chose not to return by York Factory (the annual Hudson's Bay Company sailing ship from York Factory, the *Prince Rupert*, reached London on 25 October 1849), but instead travelled via the Great Lakes and Montreal to Boston and arrived in Liverpool on 6 November 1849 (Richardson, *Arctic Searching Expedition*, 322–3).
4 The *Plover* was to act as a base ship for the other vessels on the expedition. It had to contend with terrible headwinds and "adverse currents" during its entire voyage, and did not in fact head north until late August, when the open spaces of water would have already been shrinking (Pullen, *The Pullen Expedition*, 28).
5 Whaling ships provided the easiest form of communication between expeditions and England, but they remained out only from mid-March to late August (M.J. Ross, *Polar Pioneers*, 38–9).
6 Although the first notice of the reward had accompanied Lady Franklin's 23 March 1849 broadsheet (document 3), notice of the reward was officially announced by Sir Francis Baring (1796–1866), First Lord of the Admiralty (1849–52) in Parliament on 23 April 1849 (*DNB*, 1:113; *HPD*, 1849/CIII/1188). It

did not appear in the *Gazette* until 8 March 1850, it being too late in the 1849 season for the Admiralty to ready an expedition.

7 In August 1849, the Russian Imperial Government announced its willingness to help in the search. In 1850, the Admiralty received news that a notice, "traduit en langue Russe au nombre de 300 exemplaires, sera distribuée au printems [*sic*] prochain parnis les habitans [*sic*] du littoral vers lequel se dirigent les présentes récherches" (*HCSP*, 1850/XXXV/278) [300 copies of a notice, translated into Russian, will be distributed in the coming spring to Russians inhabiting the coast that is designated as an area of search (my own translation)]. In 1851, Russia's east coast was searched for pieces of wreckage (*HCSP*, 1852/L/797–8).

8 Eleanor Franklin was at this time very busy preparing for her marriage to John Gell, and did not seem overly concerned about the British Admiralty's inability to find her missing father (Woodward, *Portrait of Jane*, 263).

Document 5

1 The Admiralty notice was announced in Parliament 23 April 1849, though it had already been offered to whalers with Lady Franklin's 23 March 1849 reward; see document 3 headnote; document 4n6.

2 Sir Francis Baring. See document 4n6.

3 According to Woodward, when Sir James Clark Ross embarked in 1848, he intended to "search both shores of Lancaster Sound and of Barrow Strait and if possible to go up Wellington Channel," eventually pushing towards Melville Island where he would meet with Sir John Richardson's overland searching expedition. Franklin was directed to Wellington Channel only if his first route, through the uncharted southwest, was impracticable (Woodward, *Portrait of Jane*, 258).

4 Captain Edward Joseph Bird (1799–1881) of the *Investigator*, Ross's second ship, was to command sledging search parties "across the ice along the west coast of Boothia and southwards" while Ross continued to sail west (Woodward, *Portrait of Jane*, 258; *Sailing Navies.com*).

5 Rae had been employed by the Hudson's Bay Company to map the shores of Melville and Simpson Peninsulas (1846–47), but his explorations were south of the region Lady Franklin refers to here. See Houston, "John Rae," 78.

6 Ross was instructed to explore Lancaster Sound, Barrow Strait, and Wellington Channel, sending sledging parties both south and west, eventually to meet up with Richardson and Rae's expedition (Savours, *Search for the North West Passage*, 189).

7 The *North Star* was to be sent to the eastern Arctic with supplies for Sir James Clark Ross's expedition; it was meant only to provide extra stores and was not itself provisioned for an extended stay in the North.

8 Dockyard Lighters were smaller, flat-bottomed boats used for unloading large ships that could not dock. The dockyards were specifically government-run harbours where Royal Navy ships were built, repaired, and outfitted.

9 Both Eleanor Franklin and John Gell were increasingly against Lady Franklin's use of family funds to search for Sir John (Woodward, *Portrait of Jane*, 263).

Document 6

1 John Middleton Clayton (1796–1856), a Whig senator from Delaware and secretary of state under President Taylor, 1849–50 (*DAB*, 8:2).

2 In his letter, written at the behest of President Taylor, Clayton expressed his country's sympathy with Lady Franklin and his government's determination that "all that the Executive Government of the United States, in the exercise of its constitutional powers can effect to meet this requisition on American enterprise, skill and bravery, will be promptly undertaken" (John Taylor to Lady Franklin, 25 April 1849, H, A/IV/N/5b).

3 A notice appeared on 22 May 1849 in the *Times* stating the American government's intentions to outfit an expedition. See Chronology.

Document 7

1 This most likely refers to the matter of document 2; see document 2n4.

2 The letters to which Lady Franklin refers are her own to President Taylor (document 4) and John Clayton's response (see document 6n2), and are not included in this section of the Hughenden collection (see document headings for exact location).

3 Sir Robert Harry Inglis (1786–1855), baronet (1820). Inglis represented Oxford University in Parliament in 1829–54 and was a Protectionist from 1847 until his retirement (*DNB*, 10:803–4). He was a powerful speaker and "exceedingly popular" in the House of Commons – a perfect representative to rouse the troops on behalf of Lady Franklin (804).

4 After Lady Franklin's appeal to President Taylor (document 4) and the enthusiastic reply, Lady Franklin, as well as some members of Parliament, rightly felt Britain's obligation to acknowledge the show of American generosity, even if the search expedition was not yet guaranteed (see documents 6, 8, 9). Disraeli did make a generous speech on behalf of Lady Franklin and the proposed

American expedition. He called Lady Franklin's letter to Taylor "worthy of a sacred cause and an heroic husband." See the *Times*, 13 June 1849, 3b.

Document 8

1 A parish southwest of London, used as a naval depot. It was the primary port where naval ships would be outfitted for journeys, including Arctic expeditions (*G*, 14).

2 Despite her personal involvement in the searches, Lady Franklin – along with the general public – got the majority of her information about Parliament's treatment of the Franklin question from the various London newspapers. Disraeli's speech appears in the *Times* of 13 June 1849, 3b.

3 Edward Jesse (1780–1868), a nature writer and deputy surveyor of royal parks and palaces (*DNB*, 10:803–4). His interest in the Franklin expedition remains undocumented. Jesse's son was a sailor, which may have increased his interest in the Franklin question. The letters sent to the *Times* by Jesse were Lady Franklin's appeal to President Taylor (document 4) and John Clayton's reply (see document 6n2).

4 The same letters that Lady Franklin sent to Disraeli on 6 June 1849 appear in the *Times*. She is most likely assuming her letters were given to Jesse by Disraeli (this would also explain their absence from the Hughenden Archives).

Document 9

1 Taylor was unable to outfit two ships for the expedition "in consequence of the want of vessels suitable to encounter the perils of a proper exploration, the lateness of the season, and the want of an appropriation by Congress to enable [him] to furnish and equip an efficient squadron for that object." In an effort to inspire the American public to do what Congress could not, Taylor had "the advertisements of reward promulgated by the British Government, and the best information [he] could obtain as to the means of finding the vessels under the command of Sir John Franklin" circulated at American ports (*HCSP*, 1850/XXXV/323). Taylor had copies of the Arctic Select Committee's 1849 opinions on the most likely routes for the discovery of Franklin (*HCSP*, 1850/XXXV/328). See Chronology.

2 There were two main reasons why Ross returned so early: his men were ill with scurvy, having been supplied with improperly cured lemon juice for the voyage; and the ice that season was unusually bad (M.J. Ross, *Polar Pioneers*, 310).

3 Ross had actually travelled in another direction than that in which he had been instructed. The route to the southwest was blocked by ice, and Ross became convinced that Franklin must have gone north towards the assumed open Polar Sea (Berton, *Arctic Grail*, 165).
4 This proposed expedition would travel around South America to enter the western Arctic in case the *Terror* and *Erebus* had in fact pushed their way through the passage. No substantial search of the western Arctic had yet been made.
5 When Congress opened on 4 January 1850, President Taylor, in response to this second appeal, submitted for the consideration of Congress "the propriety and expediency of an appropriation for fitting out an Expedition to proceed in search of the missing ships," publishing copies of his correspondence with Lady Franklin (*HCSP*, 1850/XXXV/323).

Document 10

1 Richardson had returned from his 1848–49 expedition via Boston, where he would have been able to communicate with American authorities on the subject (Richardson, *Arctic Searching Expedition*, 322–3).

Document 11

1 Captain William Alexander Baillie Hamilton, RN (1803–1881). Appointed second secretary to the Admiralty 28 January 1845, by 1850 Hamilton had become chief secretary of the Naval Department of the Admiralty at Whitehall. In this position, he frequently acted as a mediator between the Admiralty and the public and members of Parliament. A member of the Royal Geographical Society, Hamilton was also a supporter of the Franklin searches (*HBD*, 187; *POD*, 1850; *DPBK*, 351).
2 As Lady Franklin's involvement in the searches grew, so did her office space. This set of apartments was also rented for the purpose of interviewing navigators and explorers about the Arctic searches (Woodward, *Portrait of Jane*, 271).
3 Dr Robert Anstruther Goodsir appealed to Lady Franklin's "good offices" to aid him "in obtaining … an appointment of any kind in any of the Expeditions that may be sent out during the next season" (Robert Goodsir to Lady Franklin, 18 January 1850, *HCSP*, 1850/XXXV/316). Goodsir was the brother of Dr Harry D. S. Goodsir (c. 1815–48), assistant-surgeon aboard the *Erebus*, and had previously sailed with Penny in 1849 during a private summer search. See Introduction.

4 Captain Penny (see Chronology; Chronology n2) also wrote directly to the Admiralty to request a commission and to state his opinion on current searches. Although Penny had the support of some members of the navy, it was through Lady Franklin's vociferous support that he received sailing orders from the Admiralty (*HCSP*, 1850/XXXV/313–14).

5 This would have been a transcript of Kellett's summer voyage in the *Herald*, up the western coast of North America to Bering Strait. Kellett conducted three summer searches between 1848 and 1850. See Chronology.

Document 12

1 Because the Arctic was navigable for such a short time each year, organization prior to sailing was essential. From the lack of preparation made for this search, one can already see parliamentary interest flagging.

2 William Scoresby Jr (1789–1857). Scoresby travelled annually to Greenland with his father as a child, and as an adult retained his interest in Arctic matters. In 1850, he was doing volunteer clerical work while pursuing his interests in science and literature (*DNB*, 17:944–6). His book mentioned here is *An Account of Arctic Regions*, published in 1820.

Document 13

1 Lady Franklin's letter of the 13th does not exist in the official record, but her attention to "that deficiency in the Gov[t] measures" is interesting given that in April 1852 the Admiralty sent out its largest search squadron, the five-ship convoy under Sir Edward Belcher.

2 The *Isabel* had sailed under the command of Commander Inglefield on a summer voyage in 1852. Lady Franklin had offered the yacht as payment for any captain who would take it north in search of her husband. Inglefield found no trace of Franklin (Savours, *Search for the North West Passage*, 247–9).

3 The board changed hands after the election of the Aberdeen coalition government, 1852–55. This was the third Board of Admiralty with whom Lady Franklin had corresponded up to this point. See Cook and Keith, *British Historical Facts*.

4 John Pendleton Kennedy (1795–1870), secretary of the United States Navy from 26 July 1852 to 4 March 1853 under President Millard Fillmore (thirteenth US president, 1850–52). A politician and writer known for his "deep love for his fellow men," Kennedy was active in American politics as a Whig from 1820 to 1853. During his time as secretary of the US Navy, he organized four naval

expeditions, including the 1853 expedition of the *Advance*, alluded to here (*DAB*, 10:333–4). Neither a record of Kennedy's letter nor a copy of the chart mentioned here was found in *HCSP*.

5 Kane had been the surgeon aboard the first American expedition in 1850, and he planned to lead another search in 1853, commanding the *Advance* under the flag of the US Navy but with funding from Henry Grinnell, who donated the ship, and George Peabody (1795–1865), a US-born London merchant banker and philanthropist who financed the ship's refit (*DNB* online).

6 Kennedy's 4 December 1852 "Report on the Navy Department," Senate Executive Document No. 1, 32d Congress, 2d session, vol. 2, 291–326. Lady Franklin's name carried such weight in proposing motions for expeditions, she was occasionally invoked as a supporter without her knowledge.

Document 14

1 Sir James Robert George Graham (1792–1861). Graham was First Lord of the Admiralty from 1852 to 1855, under the coalition ministry of Lord Aberdeen. As First Lord, Graham's priority was achieving "administrative efficiency," which, as his opinions on the Franklin search indicate, involved not spending Admiralty funds on projects he felt to be potentially unrewarding (*DNB*, 8:328–32).

2 Graham had written a private letter to Lady Franklin on 12 January 1854 – before the news was announced in the papers – stating it was his "painful duty to announce … the decision of the Board of Admiralty, that the termination of the present financial Year, which ends on the 5th of April next the names of Sir John Franklin and of the Officers of the Erebus and Terror will be removed from the List of the Royal Navy, and that the / accumulation of Pay, which has hitherto been considered due to those Ships' Companies, will henceforth cease" (SPRI, MS 248/404/5).

3 On 14 December 1853 (8f), the *Times* announced the Admiralty's intentions to strike the names of the missing sailors from the active navy list, though it had been referring to Franklin as the "late Franklin" since 1852.

4 Embarking in 1852, Belcher's expedition was expected to stay in the Arctic for four years.

5 News of Captain McClure's discovery of a Northwest Passage had reached England, along with a few of his weakest crew members, in the autumn of 1853 (Savours, *Search for the North West Passage*, 229).

6 In a postscript to his 12 January letter, Graham amended his original assessment of the financial year ending on 5 April: "P.S. I find that the Financial

Year closes on the 31st of March, and not on the 5th of April next as stated in the above letter." He left Lady Franklin to revise her calculations based on this new date.

7 Graham stated that "after the 5th of April Claims for Pension will be received on the part of the Widows, who under the Regulations may consider themselves entitled." Relatives of the regular crew members were entitled only to their husbands' accumulated pay to 5 April.

8 Frances Woodward enigmatically records the Admiralty's treatment of Lady Franklin in 1853 as "unco-operative," but even earlier it had begun to refuse her the comprehensive aid for which she regularly appealed (Woodward, *Portrait of Jane,* 284). Lady Franklin's most recent frustration was with her ship the *Isabel*, which had been abandoned by her Admiralty crew in Chile in 1853. See Chronology; documents 16, 17.

9 Copies of Sir John Franklin's instructions were in fact given to the captains of the expeditions that went in search of him, and they explicitly stated that Franklin's first recourse was to "push to the westward without loss of time, in the latitude of about 74¼°, till you have reached the longitude of that portion of land on which Cape Walker is situated, or about 98° west. From that point we desire that every effort be used to endeavour to penetrate to the southward and westward in a course as direct toward Bhering's [*sic*] Strait as the position and extent of the ice, or the existence of land, at present unknown, may admit." Franklin's instructions admitted there was a space approximately "900 miles" between Lancaster Sound and Bering Strait that was unknown (*HCSP*, 1847–48/XLI/166). Franklin's instructions can be found in full in *HCSP*, 1847–48/XLI/163. Wallace argues in fact that Franklin's "own thinking showed a continental stress," while the belief in the open Polar Sea was more strongly felt by his superiors (Wallace, *The Navy, the Company, and Richard King,* 50).

10 Graham had pointed out to Lady Franklin that in "strictness of Law [all] Claimants to [the missing sailors' wages] ought to prove the precise Period of the Demise; because Pay can only be claimed up to the Time of the death of the deceased; and since 1846 there is no proof of their being alive. The board, however, has resolved to waive all difficulties of this kind on account of the peculiar circumstances of the Case: and the amount of Pay due will be calculated to the 5th of April of the present Year." The board had actually sought legal counsel regarding its proposal to strike the names of the missing officers and men from the active navy list. This collection of correspondence can be found in PRO, TS 18/413.

11 The board sent official notification to all relatives of the missing crews before the notice appeared in the *Gazette*.

Document 15

1 This draft on Sir John Franklin's pay is in keeping with Lady Franklin's refusal to apply for a widow's pension, which she explained to Sir James Graham in her 31 January 1854 letter. Part of this sum went to underwrite the cost incurred in Inglefield's 1852 summer expedition on the *Isabel*. See PRO, TS 18/413/2.

Document 16

1 Document 14.
2 Lady Franklin's previous count of crew members was 138 (see document 4); while 135 is repeated in later letters, this may be a typist's error in the original transcription. See document 4n2.
3 Sir Edward Belcher's expedition, which sailed in 1852.
4 See Chronology for a transcription of the *Gazette* notice.
5 In his absence, Lady Franklin received Sir John's salary from the Admiralty. She used some of this income to fund search expeditions for him (see document 15; SPRI, MS 248/212/4).
6 The current expedition was beyond the reach of the seasonal whalers and British store ships, being locked in the ice. The use of "March next" to identify the coming March is a convention. Lady Franklin means March 1854.
7 A group of Arctic navigators and experts – many of whom had been on earlier searches for the Franklin ships – were summoned by the Board of Admiralty to comprise the Arctic Select Committee, a panel formed to discuss the likely location of Franklin and the possibility of his being found. As early as 1852, however, only half of the officers questioned believed it a possibility that "Sir John Franklin, or some of his associates in adventure, do yet survive" (*HCSP*, 1852/L/217).
8 Franklin's sailing instructions can be found in full at *HCSP*, 1847–48/XLI/163; Lady Franklin discusses them at greater length later in this letter.
9 This is not an illogical assumption. Many Arctic officers gave evidence of large numbers of animals living in the high Arctic, noting that "the abundance of animal life in the region into or beyond which … [Franklin and his men would] have entered, [were] available for the extension of their original supplies" (*HCSP*, 1852/L/217). As John Rae's travelling successes proved, however, smaller numbers of experienced men were preferable to large parties, which were harder to sustain.
10 On page 4 of this letter, Lady Franklin refers to "some small vestiges of the missing ships" that were brought home in September 1850. She apparently takes these as a good sign of the crews' survival.

11 Some of the pieces of "drift pine wood" that Lady Franklin dismisses were actually pieces of British ships, found and recognized by Dr John Rae during his 1851 coastal expedition (see Chronology; Richards, *Dr John Rae*, 81).

12 Adam Beck (n.d.), Sir John Ross's Greenland Inuit interpreter, who translated for Ross on his 1850 voyage information about some Europeans who had perished in the North. Beck's account was a translation of a story he had received from a group of travelling hunters and, according to David Woodman, had nothing to do with the Franklin expedition. Beck's deposition stated that bodies of Europeans were buried beneath a cairn at North Omiak, Greenland; it was disproved in 1852 (*HCSP*, 1852/L/603). See Woodman, *Unravelling the Franklin Mystery*.

13 Reports of two abandoned ships are recorded in *HCSP*, 1852/L/941. The information given by sailors was third hand and not very recent. The alleged witnesses were interviewed, but nothing was done to look for the ships or the men in the direction from which the iceberg came.

14 Donald Beatson, a member of the Royal Geographical Society and traveller (*HCSP*, 1852/L/891–2). Beatson purchased the yacht *Isabel* specifically to search "to the north and east of Behring's Straits" for Franklin (891). When his plans to explore that area fell through, he sold the *Isabel* to Lady Franklin. Lady Franklin offered the *Isabel* to the Admiralty to use in the Arctic searches, and "made it over" in 1852 to Captain Inglefield, "on condition of his employing her during the remainder of that season in Arctic search" (SPRI, MS 248/212/5).

15 By her constant studying of maps and charts, explorers' reports, and her friendships with several Arctic navigators, Lady Franklin had become an expert in Arctic matters. From 1849 until 1860, she was convinced Franklin's ships lay in an unexplored area (see page 5 this letter). See also Introduction.

16 Captain McClure's 1850 letter to the Admiralty was transcribed in full in an editorial, titled "THE ARCTIC EXPEDITION." The article appears in the *Times*, 8 October 1853, 8a.

17 Dr Scoresby's book *The Franklin Expedition; or Considerations on Measures for the Discovery and Relief of our Absent Adventurers in the Arctic Regions* was published in 1850.

18 The 120-ton *Advance*, commanded by Dr Elisha Kent Kane on his first captaincy, left New York 31 May 1853 (*ANB*, 12:368–9; C.R. Markham, *Lands of Silence*, 298). Kane had been surgeon on the first American expedition, in 1850. The American expedition was in fact unable to reach its destination, becoming caught in the ice and eventually drifting towards Greenland.

19 Information on the searches was disseminated through the newspapers, many of which held a conservative and less-than-enthusiastic view of the considerable expense of the project. In particular, the *Times* printed scathing editorials

that did much to convince the public that the search for Franklin was a lost cause (see, for example, 13 October 1852, 4c).

20 Captain James Fitzjames (c. 1815–1848). A new captain for the 1845 expedition and second-in-command of the *Erebus*, Fitzjames had no previous Arctic experience. The journal he kept aboard the *Erebus*, which he sent home before entering the Arctic, was privately published by Mr Coningham, MP for Brighton and a very close friend (M.J. Ross, *Polar Pioneers*, 277; Smith, *Arctic Expeditions*, 386; Dickens, "The Last Leaves of a Sorrowful Book," 319).

21 Lady Franklin's testament to her husband's "determination to try both routes before his return" has disturbing implications. His desire to remain in the Arctic as long as possible, to wait for a break in the ice, may have ultimately compromised his crews' safety and kept the crew members aboard their ships long after it became necessary to abandon them. Indeed, Franklin had not turned back (at Point Turnagain) soon enough on his first Arctic overland expedition in 1821, and eleven of twenty men had perished as a result. See Franklin, *Narrative of a Journey*.

22 Austin's expedition passed west beyond the mouth of Wellington Channel before being beset by ice, while James Clark Ross was not even able to enter the pack on his 1848 expedition.

23 What Captain Penny saw in 1850 was actually a polynia, or stretch of open water found high in the Arctic, produced by tides. This discovery convinced Penny and many others that Franklin would be found only in the far north, not in the lower Arctic, and this conclusion was to misdirect the majority of searches until 1857.

24 Richardson and Rae's 1847–49 expedition. See Chronology.

25 In spite of the Admiralty's reluctance to hire him, Captain Penny was supported in this commission by Sir Francis Beaufort. Lady Franklin's support ultimately secured Penny his 1850 captaincy with the Admiralty, but it was not renewed after his return from the Arctic in 1851, largely owing to a public dispute that arose between him and Captain Austin regarding the exploration of Wellington Channel. For details of the Select Committee's findings regarding this dispute, see Holland, "The Arctic Committee of 1851"; *HCSP*, 1850/XXXV/315.

26 The "small vestiges" were the pieces of clothing and tins found by Captain Ommanney in 1850 (see Chronology).

27 That there was evidence of Franklin's "prosperous condition" during the winter of 1846 is not entirely true. The three marked graves gave at least some of the searchers concern as to the welfare of Franklin's crew; even in the extremity of Arctic conditions, three deaths in the first year was considered

unusual. Furthermore, in his 1852 deposition to the Arctic Select Committee, Captain Ommanney voiced his suspicions about the "inferior quality" of the expedition's preserved meats. The large pyramids of emptied tins left behind at the encampment – more than the expedition would have been able to consume during the first year – supported Ommanney's theory that the food was bad (in 1986, Ommanney was proved correct by Dr Owen Beattie's research). See Beattie and Geiger, *Frozen in Time*.

28 See Chronology.

29 Captain Henry Kellett, under the command of Commodore Belcher, was in charge of the western branch of the 1852 expedition. This was Kellett's fourth Arctic voyage in search of Franklin; at this time, he had been stuck in the ice two winters, conducting sledging explorations to the northwest, in the opposite direction to Franklin's ships (*DNB*, 10:1230).

30 The article appeared in the *Times*, 14 December 1853, 8f, and was printed without the Admiralty's consent. As early as 13 October 1852, however, *Times* editorials had been referring to "the late Sir John Franklin" and his "deceased" crew (4c).

31 Sir Roderick Impey Murchison (1792–1871). As president of the Geological Society from 1831, and the Geographical Society from 1843, Murchison was keenly interested in the Arctic expeditions, both pre- and post-Franklin (*DNB*, 13:1214–17). A powerful socialite, Murchison never hesitated to use his status to influence political decisions that he considered important, and he was a staunch ally of Lady Franklin in her efforts to continue the searches for Sir John.

32 Rear Admiral Sir Francis Beaufort (see document 1n3).

33 Lady Franklin had originally offered the *Isabel* to the Admiralty, if it could be used as a searching ship. The Admiralty refused but did allow Captain Inglefield a leave of absence to command it. See this document n14.

34 The *Isabel* sailed to Valparaiso, Chile, under the command of Captain Kennedy. At Valparaiso Kennedy intended to restock for the voyage north.

35 Sir Fairfax Moresby (1786–1877), admiral of the fleet (1870) and commander-in-chief of the Pacific, 1850–53 (*DNB*, 13:904–5).

36 While the ship was docked at Valparaiso, Kennedy's crew mutinied. Despite the Admiralty's earlier offer of aid to Lady Franklin and a promise of supplies, no new men were sent to run the ship, and the expedition was abandoned (Woodward, *Portrait of Jane*, 84). Barr's "'The Cold of Valparaiso'" gives an overview of the events leading up to and following the mutiny.

37 Captain Collinson's and Commander McClure's ships. Both ships entered the Arctic through the west and were supposed to meet with Austin's and Penny's

expeditions. The *Investigator* (McClure's ship) was abandoned in 1853 and later crushed by ice.

38 This is in fact untrue. When McClure and his men were rescued in 1853, many of his sailors had died, all were on the point of starvation, and half of the remaining crew could no longer stand upright due to the effects of advanced scurvy. This rumour that all McClure's men were in perfect health was propagated by McClure himself, who felt it was important for his men to walk across the Passage themselves, rather than being rescued. For the official report of the health of McClure's men, see *HCSP*, 1854–55/XXXV/75–6.

39 Sir John Ross was trapped in the ice from 1829 to 1833, during his expedition to find a Northwest Passage. Because his was a small party and Ross was adamant about anti-scorbutic consumption, his men – though on the edge of starvation – were just well enough to walk to the east coast of North America and reach help. In his published narrative of the expedition, Ross records this reaction from the whaler who rescued them: "That the mate, who commanded this boat, was as much astonished at [my identity] as he appeared to be, I do not doubt; while, with the usual blunderheadedness of men on such occasion, he assured me that I had been dead two years. I easily convinced him, however, that what ought to have been true … was a somewhat premature conclusion" (J. Ross, *Narrative of a Second Voyage*, 1: 720).

40 The account of the Russian sailors accidentally abandoned on the island of Spitzbergen was told by Sir John Richardson to the Arctic Select Committee in 1852. According to Richardson, in 1743 four men lived "six years and a quarter" (though one man died) on the island, during which they lived so prosperously that "they were able to pay for their passage home, with 2,000 pounds of deer fat, and many hides of the animals they had slain" (*HCSP*, 1852/L/238).

41 This may be Sir Edward Parry (see document 1n4), who in his book *Journal of a Second Voyage for the Discovery of a North-West Passage* (1824) asserts that "the phenomenon of two consecutive winters of extraordinary severity is one of extremely rare occurrence" (489).

42 Dr Kane's letter to the Admiralty was printed in the *Times*, 13 October 1851, 5a.

43 McClure "discovered" Banks Island, which previously had been known as Banks Land and had been only partially explored. In his efforts to sail through the Northwest Passage, McClure had tried the north as well as the south coast of Banks Island, but was frustrated by ice on both sides (Berton, *Arctic Grail*, 219–28).

44 Captain Robert Martin, captain of the last ship to be in contact with Franklin's expedition, had given a brief report of his meeting with Franklin in 1845, but was recalled to give a full deposition on 29 December 1851. It was printed in the *Times*, 3 January 1852, 5e.

45 Captain Robert Walker, master and commander of the *Union*, gave his deposition on 10 January 1852 in support of Captain Martin. It was printed in the *Times*, 14 January 1852, 8f.

46 Thomas Blanky (c. 1796–1848), acting master of the *Terror* and one of the few men on the 1845 expedition with any significant Arctic experience (M.J. Ross, *Polar Pioneers*, 278). A letter to his wife was excerpted in *Quarterly Review*, in which he stated, "Should we not be at home in the fall of 1848, or early in the spring of 1849 you may anticipate that we have made the passage, or are likely to do so; and if so, it may be from five to six years – it might be into the seventh – ere we return; and should it be so, *do not allow any person to dishearten* you on the length of our absence, but look forward with hope, that Providence will at length restore us safely to you." Before entering the Arctic, Franklin had encouraged his men to write to their families, informing them of a possible extended absence (*Quarterly Review* 92 [1852–53]: 388).

47 Commodore Belcher's instructions were suggestions by the Admiralty to direct his "special attention to the measures they now require to be adopted for at once withdrawing, if possible, the whole of the force now employed in the search for Sir J. Franklin from the Polar Seas" (*HCSP*, 1854/XLII/358).

48 The Crimean War began in 1854; the possibilities for naval action and advancement were certainly greater in the Black Sea and the Baltic, and the Admiralty was recalling all its ships to help with the war effort.

49 This "endeavour" is most likely the failed 1853 expedition of the *Isabel*. Lady Franklin's reference to her husband as one who is "too exclusively used as the representative of a corporate misfortune" is a double-edged sword; had Franklin and his crew returned to England in triumph, Franklin himself would have received the majority of the reward.

50 The Admiralty's announcement of Franklin's death caused the final split between Lady Franklin and Eleanor and John Gell. When the announcement was made, the rest of Sir John Franklin's family went into mourning, while Lady Franklin turned out in bright greens and pinks to protest what she considered a premature abandonment of her husband (Woodward, *Portrait of Jane*, 285).

Document 17

1 Document 16.

2 Sir Thomas Dyke Acland (1787–1871), Conservative MP for North Devon. Both a "politician and philanthropist," Acland was often responsible both for opening the Franklin discussion in the House of Commons and for printing many of Lady Franklin's letters (*DNB*, 1:62). He actively opposed Disraeli

in Parliament; presumably Lady Franklin hoped her husband's case would transcend party boundaries.

3 By the time the Admiralty's decision was announced in the *Gazette*, Belcher had already been sent orders to discontinue the search (see document 16n47).

4 Lord Edward Henry Stanley (1826–1893), 15th earl of Derby (1869), Conservative MP for King's Lynn (1848–69). Stanley took an interest in English popular and political culture, and would be a good advocate for Lady Franklin's interests in Parliament (*DNB*, 18:948–51). Lady Franklin did in fact write to Lord Stanley on 4 April 1854, "asking interposition to prevent the search for her husband being given up." Although Lord Stanley was willing to give support to the clause of discretionary power for an expedition's captain, he wrote in his diary that he "could not plead for the sending ~~out~~ of a new expedition" (DP 1854/43/3).

5 Disraeli was in fact unable to attend this meeting of the House, being "much indisposed," though he did mention the matter to Lord Stanley (*BDL*, VI/2649). The discussion appeared in the *Times* Wednesday 5 April 1854: 8a.

6 No copy of this letter exists in the Hughenden Archives.

Document 18

1 Franklin's sailing instructions, which Thomas Acland requested to be published in Parliament, 4 April 1854. Unfortunately, the Arctic question was not renewed after they were printed (*BDL*, VI/2649n3).

2 The *Phoenix* was meant to act only as a supply ship for Belcher's expedition.

3 Though there is no textual proof, it seems likely the captain is Leopold McClintock. Although he did return home, Captain McClintock returned to the Arctic in command of the *Fox* in 1857, financed by Lady Franklin and the last search expedition.

Document 19

1 This is Lady Franklin's first self-identification in this collection as "the widow of Sir John Franklin." This change in title, which she refused in January 1854 after notice of her husband's presumed death appeared in the *Gazette*, was the result of John Rae's discovery of relics from the missing expedition in 1854. While surveying the mainland coast, Rae had encountered a group of Inuit hunters who gave him news of a party of starved European sailors. This news had reached England in October 1854. See Chronology.

2 Sir John Richardson's letter to the editor (*Times,* 23 June, 5f) was in response to an earlier *Times* article that did not ascribe the discovery of the Northwest Passage to Franklin and his crew, but instead gave the title to Captain McClure. Richardson claims that "no dispassionate reasoner can doubt that the priority of discovery rests with the Erebus and Terror, the Investigator [McClure's ship] being at least six months later."

3 Lady Franklin in fact wrote an (eloquent as always) address to the chairman of the Arctic Select Committee to this end (see document 21).

Document 20

1 This address belongs to a Bookseller and Stationer's shop owned by the Wright family (*POD* 1850, 1856). Lady Franklin's connection with this address is not entirely clear, though stationers' addresses were frequently used by private individuals. Lady Franklin may have used the shop as an office address after she gave up her Spring Gardens apartments (Woodward 291).

2 This Arctic Select Committee had been charged with adjudicating the reward, claimed by McClure, for discovering the Northwest Passage.

3 Lady Franklin included her letter of 26 June to Palmerston (document 19), Sir John Richardson's letter to the *Times,* and Dr John Rae's letter to the *Times* in response to Richardson's, in which Rae argues that he himself should be given credit for being a "third" discoverer of a Northwest Passage (25 June 1855, 10b).

4 Lady Franklin is most likely speaking of the pieces of the *Erebus* and *Terror* that were found in the area, retrieved by Captain Collinson, and discussed by Richardson and Rae in their letters to the *Times.*

5 Inglefield announced McClure's discovery of a passage upon his own return to England in the supply steamer *Phoenix* in 1853, one year before McClure's return.

6 See document 19; Chronology.

7 The path Collinson followed during his five years on the Arctic ice also effected a "discovery of a North West Passage," though admittedly after McClure had already discovered his route. The report of McClure's triumph, however, reached England first. For his work in the Arctic, Collinson received the gold medal from the Royal Geographical Society in 1858, but he received no recognition (pecuniary or otherwise) from the Admiralty (*DNB,* 4:840).

8 Sir John Richardson wrote to Lady Franklin, thanking her for her support of his letter to the *Times,* 23 June 1855. The letter in the Hughenden collection is not a copy; it may be assumed that Disraeli did not honour Lady Franklin's wishes in this case. It can be found in H, A/IV/N/9C.

Document 21

1 William Alexander Mackinnon (1789–1870), Conservative MP (1830–52), Liberal MP (1852–65). Mackinnon acted as chairman of the Arctic Select Committee (*DNB*, 12:616–17).

2 The parallels between Franklin and McClure were quite uncanny, and Lady Franklin only touches on the similarities of their expeditions here. For instance, just as Franklin had discouraged search parties from being sent after him before his elapsed time was up, McClure wrote to the Admiralty confidently stating that "no apprehension whatever need be entertained of our safety until the autumn of 1854, as we have on board three years of all species of provisions" (quoted in Smith, *Arctic Expeditions*, 542). In fact, and in direct parallel with Franklin's experience, during McClure's first winter on the ice nearly 1,000 pounds of preserved meat were discovered to be putrid and were thrown overboard (552).

Document 22

1 In the NLC copy of this letter, some significant marginalia run alongside the first paragraphs. Although the handwriting is unidentifiable (and, being marginalia, is somewhat cramped), the style of argument is very similar to Lady Franklin's. It reads: "D[r] Rae's 'efforts' were threefold [:] 1 To listen to an inconsistent story without taking any means to verify it [;] 2 To run away as soon as possible from the only spot where he might have obtained further information [;] 3 To pocket in defiance of the rightful owners many of the relics he had brought home" (NLC, Mic.F. CC-4, no. 17703). Both Sophia Cracroft and Lady Franklin had accused Rae of keeping relics that were not rightfully his.

2 The 22 January *Gazette* notice was to announce the Admiralty's intention to give the £10,000 reward for ascertaining the fate of Franklin's expedition to Rae for his discovery of relics, if news were not heard of the expedition by the upcoming June.

3 Lady Franklin had recently undergone surgery on her throat (Woodward, *Portrait of Jane*, 291).

4 Rae explains in his report to the Admiralty that the information he records was "obtained at various times and from various sources." He submits that "none of the Esquimaux with whom [he] conversed had seen the 'whites.' Nor had they ever been at the place where the bodies were found, but had their information from those who had been there and who had seen the party when travelling" (*HLSP*, 1854–55/VII/879–97).

5 The original £20,000 offered by the Admiralty in 1849.

6 The relics Dr Rae brought home were items purchased from his Native interlocutors: crested spoons; silver pieces; and, most significantly, Sir John Franklin's Knight Commander of the Bath plaque (*HLSP*, 1854–55/VII/879–97).

7 The Back River (or Back's Fish River), surveyed by Sir George Back (1796–1878) during his unsuccessful attempt to rescue Sir John Ross in 1833 (*DNB*, 1:790–2). Back, part of Franklin's 1819–22 expedition and possibly the sole reason the expedition survived, was not an ally of Franklin's and disapproved of his being put in command of the 1845 voyage.

8 In 1855, James Anderson and James Stewart of the Hudson's Bay Company were sent to verify Rae's claims, but they were unable to visit King William Island. See Chronology.

9 Anderson and Stewart's expedition.

10 Even the *Times*, normally discouraging of any further public expenditure on the Franklin search, urged the government to conduct one last search of the area identified in Rae's report. See 24 October 1854, 6c.

11 Although Anderson and Stewart were unable to communicate verbally with the Natives they met, they were able to speak through signs: the Native women, whom Anderson termed "very intelligent," all indicated by rubbing their bellies and biting their arms the truth of Rae's report (*HLSP*, 1856/X/333–7; RGS, 1856:21).

12 The third clause of the 1850 reward was £10,000 to the person who "ascertained the fate" of the missing sailors.

13 Penny (as well as several other officers and scientists) had in fact written to the Admiralty in an attempt to claim the £10,000 that was eventually awarded to Rae (*HCSP*, 1856/XLI/167).

14 The *Isabel* was eventually declared unseaworthy and was not used again for Arctic service (Woodward, *Portrait of Jane*, 84).

15 Sir Charles Wood (1800–1885), First Viscount Halifax, Liberal MP for Halifax, 1832–64, and First Lord of the Admiralty, 1855–58 (*DNB*, 21:824–5). Wood was ordered by Prince Albert to extend an offer of support to Lady Franklin should she decide to equip a privately funded expedition (Woodward, *Portrait of Jane*, 290).

16 Ruling against Lady Franklin's claim, the Select Committee adjudicating the reward to McClure argued that it did not consider Lady Franklin's claim only because it could not be proven. See Introduction.

17 The Admiralty had in fact already drawn up a draft announcement of Rae's reward, in anticipation of the expected results of Anderson and Stewart's expedition. See *HCSP*, 1856/XLI/167.

Document 23

1 Document 22.

2 The official announcement of Rae's reward was printed in the *Gazette* of 24 June.

3 The "Memorial to the Right Hon. Viscount Palmerston, M.P., G.C.B." was presented 5 June, 1856. The petition attempted to convince Palmerston of the "desirableness of sending out an Expedition to satisfy the honor of our country, and clear up a mystery which has excited the sympathy of the civilized world." The request was "supported by many persons well versed in Arctic surveys, who, seeing that the proposed Expedition is to be directed *to one limited area* only, are of opinion that the object is attainable, and with little risk." Eminent Arctic experts and scientists who had been involved in the search and who had signed the petition included Francis Beaufort, Roderick Murchison, Lord Wrottesley, Edward Sabine, Richard Collinson, and William Hamilton. The list of absent captains who had expressed a desire to verify Rae's claims included nearly every British captain who had been involved in the search before Rae's discovery and had thus endured the earlier risk of speculative exploration. For a full transcript of the petition, see Appendix 2 in McClintock's *The Voyage of the 'Fox.'*

4 This is most likely Captain McClintock.

5 Captain Rochfort Maguire (d. 1867). Captain Maguire acted as first lieutenant under Kellett aboard the *Herald* during its six-and-a-half-year Pacific voyage and its three summer expeditions to the western Arctic (1848–50) (NMMC, II/ii/775).

6 This was in fact highly unlikely; Collinson's voyage to the land adjacent to King William Island took no less than two years during his 1850–55 expedition. See Chronology.

7 Lady Franklin included a copy of a letter from Maguire to Collinson of 28 June 1856, which stated that "a vessel may almost count on the passage in one season" to Point Barrow (*HCSP*, 1857–58/LX/37).

8 For details of Anderson's opinions regarding a new search, see document 21.

9 Captain George Henry Richards had been captain of the *Assistance* in Belcher's 1852 expedition.

Document 24

1 A copy of the memorial is printed in the same edition of F.L. McClintock, *The Voyage of the 'Fox,'* Appendix 2, 329–33. See also document 23; document 23n3.

2 Document 22.

3 There is in fact no evidence in parliamentary records that the Admiralty was discussing the possibility of another expedition. Lady Franklin is rather referring to a question put to Lord Stanley regarding the Admiralty's consideration of the 1856 memorial and the possibility of another expedition. Stanley's reply was guarded at best: he promised that Sir Charles Wood would "take into his most serious consideration … whether it was desirable that any further expedition should be sent out," but expressed doubt as to the feasibility of such an expense. Lady Franklin's interpretation here is liberal. For a full transcription of the debate, see *HPD*, 18 July 1856: 1008–13. Edward John Stanley (1802–1869), first Baron Eddisbury of Winnington (1848), second Baron Stanley of Alderley (1850), held a number of posts in Parliament, from paymaster general to principal whip of the Whig party. In 1855–58, Lord Stanley was president of the Board of Trade (*DNB*, 18:951–3). On 18 July 1856 (the date of the parliamentary discussion Lady Franklin cites), Lord Stanley spoke on Sir Charles Wood's behalf (*HPD*, 18 July 1856: 1013).

4 Document 23.

5 Sir John Wrottesley, second Baron Wrottesley (1798–1867), lifelong member of the Society for Diffusing Useful Knowledge and a member of the Royal Society since 1841. Wrottesley supported several scientific enterprises; outside of personal interest, his connection with the Franklin expedition is unexplained (*DNB*, 21:1082–3).

6 This conversation was actually an unofficial one; it does not appear in any records of the parliamentary debates or in Admiralty papers, and so would have very little weight with the government. Stanley was most likely expressing his personal opinion to Wrottesley, as his guarded promise on behalf of the Admiralty transcribed in the official record suggests that the government considered another expedition superfluous.

7 The *Resolute*, one of the four ships Belcher abandoned in 1854. See Chronology.

8 The *Resolute* was purchased by the US government and returned to England "as a token of the deep interest felt by the United States in the service in which that ship was engaged at the time she was abandoned" (*HCSP*, 1857–58/LX/41). For the full transcription of the US Resolution of Congress, see *HCSP*, 1857–58/LX/43–4.

9 Henry Grinnell (see Chronology).

10 The property inheritance, bequeathed by an uncle of James Walter Fairholme (1821–1848), a lieutenant aboard the *Erebus*, was contested by a cousin. Ultimately, the court ruled in the cousin's favour, as James Fairholme had died in 1848 and so had predeceased his uncle, who died in 1853. See the *Times*, 26 January 1857, 12b.

11 This is a paraphrase from Sir John Richardson's biography of Sir John Franklin (and the Franklin expedition) in *Encyclopedia Britannica*, 8th edition.

12 It was well known even in 1857 that written material did not easily survive the Arctic climate. James Anderson states in his 1855 report to the Admiralty that the Arctic wind was powerful enough to tear pages out of a leather-bound book, and the 1852–53 issue of the *Quarterly Review* describes the fate of Dr Rae's books from his 1846 Arctic travels thus: "[T]he leaves were fast frozen together, the damp from the walls having got into them before the frost set in; and every article bound with brass or silver [i.e., official ships' logbooks and other bound texts] burst its fastenings" (*Quarterly Review* 92:392).

13 Sir Charles Beaumont Phipps (1801–1866), court official and colonel unattached. Phipps was a favourite of the royal family: in 1846, he was made equerry to Queen Victoria; in 1847, he became private secretary to the prince consort and the court treasurer (*DNB*, 15:1114). Phipps likely acted as Prince Albert's messenger to Sir Charles Wood in this instance.

Document 25

1 Richard Monckton Milnes (1809–85), MP for Pontefract 1837–46 as a Tory, and from 1846 to 1863 as a Liberal. Though known as generous and "always ready to offer his help," Milnes was "eminently a dilettante; while his interests were wide, he shirked the trouble necessary for judgments other than superficial" (*BDL*, III/l750n6; *DNB*, 13:468).

2 Sir Charles Napier (1786–1860), admiral (1858) and MP for Southwark, 1855–60. After a falling out with the Admiralty in 1854, Napier "in and out of parliament devoted himself to denouncing ... the board of Admiralty" (*DNB*, 14:38–45). The Admiralty's handling of the Franklin search was a perfect point for criticism. On 24 February 1857, Napier moved to reopen the discussion of the Arctic searches. His motion was later withdrawn. See the *Times*, 25 February 1857, 8e.

3 The expedition by the Hudson's Bay Company, under James Anderson and James Stewart, 1855 (see Chronology).

4 Lady Franklin's series of letters regarding this final search received either equivocal replies or no replies at all. See documents 22, 23, 24.

5 See document 24.

6 Lieutenant Hartstene (see Chronology).

7 Lady Franklin most likely means the communication given to her by Colonel Phipps (see document 24).

Document 26

1 Sir Charles Wood had announced that "though the Government were not insensible to the appeal very naturally made by Lady Franklin for pursuing the search," there would be no more government expeditions if there was no chance of recovering living crew members. See *HPD*, 24 February 1857: 1282.

2 The resolution of Congress (from which this quotation is taken) can be found in full in *HCSP*, 1857–58/LX/43–4.

3 George Mifflin Dallas (1792–1864), lawyer and American minister to Great Britain, 1856–61. Dallas was exceedingly popular in Britain, and his diplomacy was central to maintaining good relations between Britain and the United States during the Crimean War (*ANB*, 6:31–3). Dallas's reference (quoted here) to the *Resolute* is from a speech he had given to the Royal Geographical Society.

4 This would have been the *Isabel*, under Captain Kennedy (1853). This is a much different interpretation than Lady Franklin previously gave to the Admiralty's support for this expedition. See documents 16, 17.

Document 27

1 This letter does not exist in the official record or correspondence regarding the *Resolute* in *HCSP*.

2 There is no record of this quote in official papers; it is most likely an extract from Lord Wrottesley's letter to Lady Franklin and a transcript of his unofficial conversation with Lord Stanley of Alderley (see document 24n6).

3 This representation also does not appear in the official record of correspondence regarding the *Resolute*.

4 Lady Franklin's private expedition was partially funded by public subscription; individuals donated from 2*s*. 6*d*. to £500. The largest contribution was given anonymously by Captain Allen William Young (later Sir) (1827–1915), sailing-master of the *Fox*. Allen also refused any salary for his labour. He was knighted in 1877 (*DNB*, suppl. 3: 599–600). Many other substantial contributions of up to £150 were given by officers involved in the earlier searches for Franklin and by relatives of the missing sailors. For a complete list of subscribers, see F.L. McClintock, *The Voyage of the 'Fox,'* Appendix 5, 400–3.

Document 28

1 Lady Franklin and Sophia Cracroft's residence, 1859–74 (Woodward, *Portrait of Jane*, 304).

2 Sarah Disraeli (1802–1859). Elder sister of Benjamin Disraeli, and old family friend of Lady Franklin (*BDL*, I/10n1).

3 Many of Lady Franklin's personal relationships suffered during her efforts to motivate the Admiralty and government to search for her husband. The volume that she sends to Disraeli is a copy of Captain McClintock's *The Voyage of the 'Fox': A Narrative of the Discovery of the Fate of Sir John Franklin and his Companions*, published late in 1859. McClintock dedicates his book to Lady Franklin, saying that it is to her "devotion and self-sacrifice [that] the world is indebted for the deeply interesting revelation unfolded by the voyage of the 'Fox'" (F.L. McClintock, *The Voyage of the 'Fox,'* iii).

Document 29

1 On 2 March 1860, Baring had introduced the possibility of voting public funds towards a reward for McClintock and his crew, the £10,000 officially allocated for ascertaining the fate of Franklin and his men having already been awarded to Rae. Lord Palmerston expressed interest in hearing the feelings of the House on the matter, and the motion was unanimously passed. Disraeli spoke briefly on the subject, praising both McClintock and Lady Franklin for their efforts: "No on can have read the narrative of Sir Leopold McClintock, or marked the progress of his singular expedition, without feeling that their adventure will form a memorable chapter in the heroic annals of the mariners of this country. With regard to that noble lady through whose energy and inspiration that expedition was so gallantly undertaken and so successfully accomplished, I feel, as I am sure we all must feel, the liveliest admiration and the warmest sympathy for the rare intelligence, the indomitable perseverance, the womanly, the high and enduring spirit, with which Lady Franklin urged on the prosecution of the search for her brave husband – they will make her name illustrious among the daughters of England" (*HPD*, 2 March 1860: 2157).

BIBLIOGRAPHY

UNPUBLISHED MATERIAL

Derby Papers, 15th Earl, Liverpool City Libraries
Lord Stanley Diary, DBP, Box 43: 3 (1854)
Hughenden Papers, Bodleian Library, Oxford
Lady Franklin Letters, H, A/4/N/2-15
National Library of Canada, Ottawa
Archives
Public Record Office, Kew
Correspondence Relating to Sir John Franklin, TS 18/413
Scott Polar Research Institute, Cambridge University
Lady Franklin Letters, MS 248/212/1-101
Lady Franklin Correspondence, MS 248/404:5

GOVERNMENT DOCUMENTS

First titles of documents, volume numbers, and beginning page numbers have been given.

House of Commons Sessional Papers

1847–48/LXI/163. "Copies of Instructions to Captain Sir *John Franklin*, R. N., in reference to the Arctic Expedition of 1845."

1849/XXXII/217. "Correspondence with H.M. *Chargé d'Affaires* at Washington, respecting Intention of Govt. of United States to send Ships-of-War in Search of Sir J. Franklin's Expedition to Arctic Seas."

1850/XXXV/175. "Copies of any Reports or Statements from the Officers employed in the Arctic Expeditions, or from any other Persons, which have been laid before the Lords Commissioners of the Admiralty, in respect to the Resumption of the Search for Sir *John Franklin*'s Expedition."

1851/XXXIII/195. "Correspondence and Papers in relation to Arctic Expeditions in Search of Sir J. Franklin, with Information relating to United States' Expedition."
1852/L/1. "Report of the Committee appointed by the Lords Commissioners of the Admiralty to enquire into and report on the Recent Arctic Expeditions in Search of Sir John Franklin, together with the Minutes of Evidence taken before the Committee, and Papers Connected with the Subject."
1852/L/669. "Further Correspondence and Procs. connected with Arctic Expedition."
1852/L/939. "Communications between Admiralty and Public Authorities in reference to certain Vessels observed on Iceberg in North Atlantic, 1851."
1854/XLII/339. "Copy of Letter addressed by Lady *Franklin* to the Lords Commissioners of the Admiralty, dated the 24th day of February 1854, in reference to their Lordships' Announcement in the London Gazette of the 20th day of January 1854, respecting the Officers and Crews of Her Majesty's Ships 'Erebus' and 'Terror;' and of a Copy of such Notice."
1854/XLII/353. "Copies of any Instructions which either have been Issued, or hereafter may be Issued, during the present Season, to the Commanders of Her Majesty's Ships now engaged in the Arctic Regions in the Search for Sir *John Franklin*'s Expedition."
1854–55/VII/3. "Report from the Select Committee on Arctic Expedition; together with the Proceedings of the Committee, Minutes of Evidence, and Appendix."
1854–55/XXXV/1. "Further Papers relative to the recent Arctic Expeditions in Search of Sir *John Franklin*, and the Crews of H.M.S. '*Erebus*' and '*Terror*.'"
1856/XLI/111. "Further Papers relative to the Recent Arctic Expeditions in Search of Sir J. Franklin, and the crews of Her Majesty's Ships 'Erebus' and 'Terror.'"
1857–58/LX/11. "Correspondence respecting H.M.S. 'Resolute,' and the Arctic Expedition."

House of Lords Sessional Papers

1854–55/VII/3. "Further Papers relative to the recent Arctic Expeditions in search of Sir John Franklin, and the Crews of H.M.S. 'Erebus' and 'Terror.'"
1856/X/325. "Proceedings of the Hudson's Bay Company Arctic Expedition, sent down the Great Fish River, under Command of Messrs. Anderson and Stewart, to investigate the Traces of the Party under Sir John Franklin found by Dr. Rae."

1856/x/355. "Correspondence relating to the adjudication of the sum of £10,000 as a reward for ascertaining the fate of the crews of Her Majesty's ships 'Erebus' and 'Terror.'"

PUBLISHED SOURCES

Amy, Roger, et al. "The Last Franklin Expedition: Report of a Postmortem Examination of a Crew Member." *Canadian Medical Association Journal* 135 (1986): 115–17.

Athenaeum 1048 (1847); 1528 (1857); 1665 (1859); 1678 (1859).

Barr, William. "'The Cold of Valparaiso': The Disintegration of William Kennedy's Second Franklin Search Expedition, 1853–1854." *Polar Record* 34, no. 190 (1998): 203–18.

–, ed. *Searching for Franklin: The Land Arctic Searching Expedition. James Anderson's and James Stewart's Expedition via the Back River, 1855*. London: Hakluyt Society, 1999.

Barrow, John. *A Chronological History of Voyages into the Arctic Regions, Undertaken Chiefly for the Purpose of Discovering a North-East, North-West or Polar Passage Between the Atlantic and Pacific*. 1818. New York: Barnes and Noble, 1971.

Beattie, Owen. "Elevated Bone Lead Levels in a Crewman from the Last Arctic Expedition of Sir John Franklin (1845–1848)." In *The Franklin Era in Canadian Arctic History 1845–1859*, edited by Patricia D. Sutherland, 141–8. Ottawa: National Museums of Canada, 1985.

Beattie, Owen, and John Geiger. *Frozen in Time: Unlocking the Secrets of the Franklin Expedition*. Saskatoon: Western Producer Prairie Books, 1987.

Behrisch, Erika. "On the Trail of an Arctic Tale: Tracing Sir John Franklin in Charles Dickens and Wilkie Collins's *The Frozen Deep*." In *Storytelling: Interdisciplinary and Intercultural Perspectives*, edited by Irene Blayer and Monica Sanchez, 58–71. New York: Peter Lang, 2002.

– "'Far as the eye can reach': Scientific Exploration and Explorers' Poetry in the Arctic, 1832–1852." *Victorian Poetry* 41, no. 1 (2003): 73–92.

Benjamin Disraeli Letters. Edited by M.G. Wiebe, Mary Millar, et al. 7 vols. Toronto: University of Toronto Press, 1982– .

Berton, Pierre. *The Arctic Grail: The Quest for the North West Passage and the North Pole, 1818–1909*. Toronto: McClelland and Stewart, 1988.

Bloom, Lisa. *Gender on Ice: American Ideologies of Polar Expeditions*. Minneapolis: University of Minnesota Press, 1993.

Boase, Frederic. *Modern English Biography*. 3 vols. 1897. London: Frank Cass and Company, 1965.

Brake, Laurel. "Writing, Cultural Production, and the Periodical Press in the Nineteenth Century." In *Writing and Victorianism*, edited by J.B. Bullen, 54–72. London: Longman Press, 1997.

Campbell, William Wilfred. "Unabsolved." In *Beyond the Hills of Dream*, 95–106. Boston: Houghton Mifflin, 1899.

Christ, Carol T. "'The Hero as Man of Letters': Masculinity and Victorian Nonfiction Prose." In *Victorian Sages and Cultural Discourse: Renegotiating Gender and Power*, edited by Thaïs E. Morgan, 19–31. New Brunswick, NJ: Rutgers University Press, 1990.

Cook, Chris, and Brendan Keith. *British Historical Facts*. London: Macmillan Press, 1975.

Cook, Ramsay, ed. *Dictionary of Canadian Biography*. 14 vols. Toronto: University of Toronto Press, 1994.

Cooke, Alan, and Clive Holland. *The Exploration of Northern Canada: 500 to 1920, A Chronology*. Toronto: Arctic History Press, 1978.

Cookman, Scott. *Ice Blink: The Tragic Fate of Sir John Franklin's Lost Polar Expedition*. New York: Wiley, 2000.

Corner, George W. *Doctor Kane of the Arctic Seas*. Philadelphia: Temple University Press, 1972.

Cracroft, Sophia. *Lady Franklin Visits Sitka, Alaska 1870: The Journal of Sophia Cracroft, Sir John Franklin's Niece*. Edited by R.N. DeArmond. Anchorage: Alaska Historical Society, 1981.

Crossing the Line Films. *Finding Franklin*. London: 2005.

Cruwys, Liz. "Edwin Jesse De Haven: The First US Arctic Explorer." *Polar Record* 28, no. 166 (1992): 205–12.

Cunningham, Peter. *Hand Book of London*. London: John Murray, 1850.

Cyriax, Richard J. *Sir John Franklin's Last Arctic Expedition; A Chapter in the History of the Royal Navy*. London: Methuen, 1939.

David, Robert G. *The Arctic in the British Imagination, 1818–1914*. Manchester: Manchester University Press, 2000.

DeArmond, R.N. "Introduction." In Cracroft, *Lady Franklin Visits Sitka*, xiii–xviii.

Debrett's Peerage, Baronetage and Knightage 1893. London: Dean and Son Ltd, 1893.

Delgado, James P. *Across the Top of the World: The Quest for the Northwest Passage*. Vancouver: Douglas and McIntyre, 1999.

Dickens, Charles. "Dr Rae's Report." *Household Words* 249 (1854): 457–9.

– "The Lost Arctic Voyagers." *Household Words* 245 (1854): 361–5.

– "The Lost Arctic Voyagers." *Household Words* 246 (1854): 385–92.

– "The Lost Arctic Voyagers." *Household Words* 248 (1854): 433–7.

– "The Last Leaves of a Sorrowful Book." *All the Year Round* 1, no. 26 (1859): 318–23.

Dictionary of American Biography. Edited by Dumas Malone. 22 vols. New York: Charles Scribner's Sons, 1932.

Dictionary of National Biography. Edited by Sir Leslie Stephen and Sir Sidney Lee. 22 vols. Oxford: Oxford University Press, 1917. *DNB Online*: www.oxforddnb.com.

Dramin, Edward. "'Work of Noble Note': Tennyson's 'Ulysses' and Victorian Heroic Ideals." *Victorian Literature and Culture* 20 (1992): 117–39.

Encyclopædia Britannica. 29 vols. 11th ed.

"Fate of the Franklin Expedition." *Leisure Hour* 417 (1859): 812–16.

Fleming, Fergus. *Barrow's Boys.* London: Granta Books, 1998.

"Forlorn Hope." *Fraser's Magazine* 49 (1854): 633–45.

Franklin, John. *Narrative of a Journey to the Shores of the Polar Sea in the Years 1819, 1820, 1821, 1822.* London: John Murray, 1823.

Garraty, John A., and Mark C. Carnes, eds. *American National Biography.* 24 vols. New York: Oxford University Press, 1999.

Gazette (London), 8 March 1850; 20 January 1854; 22 January 1855; 24 June 1855.

Gazetteer of the British Isles.

Gibson, William. "Sir John Franklin's Last Voyage." *Beaver* 268, no. 1 (1937): 44–75.

Goodsir, Robert A. *An Arctic Voyage to Baffin's Bay and Lancaster Sound in Search of Friends with Sir John Franklin.* London: J. Van Voorst, 1850.

Hansard's Parliamentary Debates. 3rd Series. London: G. Woodfall and Son.

Harries, Elizabeth. "Out in Left Field: Charlotte Smith, Bourdieu and the Public Sphere." *Modern Language Quarterly* 1997:457–73.

Haydn, Joseph. *Joseph Haydn's The Book of Dignities.* 3rd ed. 1894. Baltimore: Genealogical Publishing Company, 1970.

Holland, Clive. "The Arctic Committee of 1851: A Background Study." *Polar Record* 20, no. 124 (1980): 3–17.

Houston, C. Stuart. "Dr. John Rae: The Most Efficient Arctic Explorer." *Annals RCPSC* 20, no. 3 (1987): 225–8.

– "John Rae (1813–1893)." *Arctic* 40 (1987): 78–9.

Journal of the Royal Geographical Society. London: John Murray.

Kennedy, John Pendleton. "Report on the Navy Department." Senate Executive Document No. 1, 32nd Congress, 2nd session, vol. 2, 291–326.

London Post Office Directory, 1850, 1856, 1860.

Loomis, Chauncey. "The Arctic Sublime." In *Nature and the Victorian Imagination,* edited by U.C. Knoepflmacher and G.B. Tennyson, 95–112. Berkeley: University of California Press, 1977.

Lyon, George. *A Brief Narrative of an Unsuccessful Attempt to Reach Repulse Bay, through Sir Thomas Rowe's "Welcome," in His Majesty's Ship Griper, in the year* MDCCCIXV. London: John Murray, 1825.

McClintock, Anne. *Imperial Leather: Race, Gender and Sexuality in the Colonial Contest.* New York: Routledge, 1995.

McClintock, F. Leopold. *The Voyage of the 'Fox': Narrative of the Discovery of the Fate of Sir John Franklin and his Companions.* London: John Murray, 1859.

McGoogan, Ken. *Fatal Passage: The Untold Story of John Rae, the Arctic Adventurer who Discovered the Fate of Franklin.* Toronto: HarperCollins, 2001.

– *Lady Franklin's Revenge: A True Story of Ambition, Obsession and the Remaking of Arctic History.* Toronto: HarperCollins, 2005.

MacLaren, I.S. "The Poetry of the 'New Georgia Gazette' or 'Winter Chronicle' 1819–1820." *Canadian Poetry* 30 (1992): 41–73.

– "From Exploration to Publication: The Evolution of a 19th-Century Arctic Narrative." *Arctic* 47, no. 1 (1994): 43–53.

Markham, A.H. *Life of Sir John Franklin and the North-west Passage.* London: George Philip and Son, 1891.

Markham, Clements R., KCB, FRS. *The Lands of Silence: A History of Arctic and Antarctic Exploration.* Cambridge, UK: Cambridge University Press, 1921.

"Memorial to the Right Hon. Viscount Palmerston, M.P., G.C.B." In F.L. McClintock. *The Voyage of the 'Fox,'* Appendix 2, 329–33.

Middleton, Dorothy. "Franklin [*née* Griffin], Jane, Lady Franklin (1792–1875)." 2004. *DNB Online.* www.oxforddnb.com/view/article/10089.

[Morley, Henry]. "Unspotted Snow." *Household Words* 190 (1853): 241–46.

National Maritime Museum Catalogue of the Library. 2 vols. London: Her Majesty's Stationery Office, 1969.

Neatby, Leslie H. *The Search for Franklin.* New York: Walker and Company, 1970.

O'Neill, Patrick. "Theatre in the North: Staging Practices of the British Navy in the Canadian Arctic." *Dalhousie Review* 74, no. 3 (1994): 356–84.

Osborn, Sherard. *Stray Leaves from An Arctic Journal; or, Eighteen Months in the Polar Regions, in Search of Sir John Franklin's Expedition, in the Years 1850–51.* London: Longman, Brown, Green, and Longmans, 1852.

Owen, Roderic. *The Fate of Franklin.* London: Hutchinson, 1978.

Parkinson, Edward. "'All Well': Narrating the Third Franklin Expedition." In *Echoing Silence*, edited by John Moss, 43–52. Ottawa: Ottawa University Press, 1996.

Parry, William Edward. *Journal of a Second Voyage for the Discovery of a North-West Passage.* 1824. New York: Greenwood Press, 1969.
Pullen, H.F. *The Pullen Expedition.* Toronto: Arctic History Press, 1979.
Quarterly Review 92 (1852–53).
Rae, John. "The Lost Arctic Voyagers." *Household Words* 248 (1854): 433–7.
Rawnsley, W.F. *The Life, Diaries and Correspondence of Jane Lady Franklin.* London: Erskine MacDonald, 1923.
Richards, Robert L. *Dr John Rae.* Whitby, UK: Caedmon of Whitby, 1985.
Richardson, Sir John. *Arctic Searching Expedition: A Journal of a Boat-voyage through Rupert's Land and the Arctic Sea, in Search of the Discovery Ships under Command of Sir John Franklin.* New York: Harper and Brothers, 1852.
Ross, John. *Narrative of a Second Voyage in Search of a North-West passage, and of a Residence in the Arctic Regions During the Years 1829, 1830, 1831, 1832, 1833.* 2 vols. London: A.W. Webster, 1835.
Ross, M.J. *Polar Pioneers.* Montreal and Kingston: McGill-Queen's University Press, 1994.
Ross, W. Gillies. *This Distant and Unsurveyed Country: A Woman's Winter at Baffin Island, 1857–58.* Montreal and Kingston: McGill-Queen's University Press, 1997.
– "The Arctic Council of 1851: Fact or Fancy?" *Polar Record* 40 (2004): 135–41.
Sailing Navies 1650–1850. Sailing Navies.com. www.sailingnavies.com.
Savours, Ann. *The Search for the North West Passage.* New York: St Martin's Press, 1999.
Smith, D. Murray. *Arctic Expeditions: British and Foreign From the Earliest Period to the Present Time.* London: Thomas C. Jack, 1880.
Spufford, Francis. *I May Be Some Time: Ice and the English Imagination.* London: Faber & Faber, 1996.
Steele, David. "Temple, Henry John, third Viscount Palmerston." 2004. *DNB Online*: www.oxforddnb.com/view/article/27112.
Stone, Ian. "An Episode in the Franklin Search: The *Prince Albert* Expedition, 1850." Part 1. *Polar Record* 29, no. 169 (1993): 127–42.
– "The Franklin Search in Parliament." *Polar Record* 32, no. 182 (1996): 209–16.
Struzik, Edward. *Northwest Passage.* Toronto: Key Porter Books, 1991.
Thomson, George Malcolm. *The North-West Passage.* London: Secker and Warburg, 1975.
Times (London), 22 May 1849; 13 June 1849; 13 October 1851; 3 January 1852; 14 January 1852; 13 October 1852; 8 October 1853; 14 December 1853; 4 April 1854; 23 October 1854; 24 October 1854; 23 June 1855; 25 June 1855; 26 January 1857; 25 February 1857; 15 April 1857; 29 April 1857; 30 April 1857; 3 March 1860.

Traill, Henry Duff. *The Life of Sir John Franklin, R.N.* London: John Murray, 1896.

Tyrell, J.B. "A Story of a Franklin Search Expedition." *Transactions of the Canadian Institute* 8 (1908–09): 393–402.

Vancouver Maritime Museum. http://vancouvermaritimemuseum.com/modules/vmmuseum/treasures/?artifactid=181/ (accessed 15 November 2007).

W., C.R. "Franklin's Fate, and the Voyage of the 'Fox.'" *Fraser's Magazine* 61 (1860): 221–7.

Wallace, Hugh N. *The Navy, the Company, and Richard King: British Explorations in the Canadian Arctic, 1829–1860.* Montreal and Kingston: McGill-Queen's University Press, 1980.

Wemyss Reid, T., ed. *The Life, Letters and Friendships of Richard Monckton Milnes First Lord Houghton.* 2 vols. London: Cassell & Company, 1890.

Wiebe, Rudy. *Playing Dead: A Contemplation Concerning the Arctic.* Edmonton: NeWest Publishers, 1989.

Williams, Glyndwr. *The British Search for the Northwest Passage in the Eighteenth Century.* London: Longmans, 1962.

Wilson, John. *North with Franklin: The Lost Journals of James Fitzjames.* Markham: Fitzhenry and Whiteside, 1999.

Woodman, David C. *Unravelling the Franklin Mystery: Inuit Testimony.* Montreal and Kingston: McGill-Queen's University Press, 1991.

Woodward, Frances J. *Portrait of Jane: A Life of Lady Franklin.* London: Hodder and Stoughton, 1951.

Wright, Noel. *Quest for Franklin.* London: Heinemann, 1959.

INDEX

Numbers in bold indicate where a brief biography of the subject can be found.